THE MOLECULAR CHARACTERIZATION OF THYROID STIMULATING HORMONE (TSH) IN THYROID GLAND TUMORS BY RADIOBINDING ASSAY TECNIQUES

Prof. Dr. Sami AL-Mudhaffar

Suha Hanoun Hassan

Abbreviations

^{125}I-Anti tTSH Ab	Iodine labeled anti total TSH antibody
A	Absorbance
AAb	Auto antibodies
Ab	Antibody
ACTH	Adrenocorticotrophic Hormone
Ag	Antigen
BSA	Bovine Serum Albumin
cAMP	Cyclic Adenosine monophosphate
CEA	Carcino-Embryonic Antigen
cpm	Counts per minute
CRH	Corticotropin Releasing Hormone
CT	Computerized Tomography
DIT	Diiodotyrosine
DNA	Deoxyribonuceic Acid
FCDC	Follicular cell derived carcinoma
FSH	Follicular stimulating hormone
FTC	Follicular thyroid carcinoma
hCG	human chorionic gonadotrophin
hTR	Human thyroid hormone receptor
ICT	Immunoreactive calcitonin
IR	Infrared
IRMA	Immunoradiometric Asaay
K	Constant, equilibrium
LH	Lutenizing Hormone
M	Molar (concentration)
Max	Maximum
Mg	Miligram
MIT	Monoiodotyrosine
MRI	Magnetic Resonance Image

MTC	Medullary Thyroid Carcinoma
N.M.R.	Nuclear Magnetic Resonance
PTC	Papillary Thyroid Carcinoma
RAI	Radioactive Iodine
RIA	Radioimmuno assay
$rL\text{-}T_3$	Reverse L-triiodothyronine
RNA	Ribonucleic acid
RRA	Radio active iodine ablation
SD	Standard deviation
TBA	Thyroxin Binding Albumin
TBG	Thyroxin Binding Globulin
TBPA	Thyroxin Binding Pre-Albumin
Tg	Thyroglobulin
$TL\text{-}T_3$	Total L-triiodothyronine
$TL\text{-}T_4$	Total L-thyroxine
tTSH	Total TSH
TNM	Tumor, Node, Metastasis, Staging System
TRH	Thyrotropin Relasing Hormone
TSH	Thyroid Stimulating Hormone
TSH-R	TSH-Receptor
UV	Ultraviolet
WHO	World Health Organization
ε	Absorption coefficient
μ	Micro (10^{-6} x)
μg	Microgram
μu	Microunit

Summary

1. An Immunoradiometric assay (IRMA) was applied in determination of thyroid stimulating hormone (TSH) levels in sera and thyroid tissue homogenates of patients with benign and malignant thyroid tumors. In addition, the TSH level was also determined in sera of closely matched (by both sex and age) normal persons to be used as control group. On the other hand, Radioimmuno assay (RIA) was applied in determination of both L-thyroxine (L-T$_4$) and L-triiodothyronine (L-T$_3$) levels in sera of the benign and malignant thyroid tumor patients in comparison to that of the closely matched controls.

2. The characteristics of the binding of ^{125}I-anti total TSH antibody with TSH in benign and malignant thyroid tissue homogenates were also investigated. Different factors affecting this binding were extensively studied such as pH, time, temperature, concentration of salts, concentration of antibody and concentration of antigen.

3. The kinetic parameters such as B_{max}, k_a, k_d of binding of ^{125}I-anti tTSH Ab with TSH in benign and malignant thyroid tissue homogenates were determined at five different temperatures, while $k_{obs.}$, k_{+1}, k_{-1} and $t_{1/2}$ of the binding of ^{125}I-anti tTSH Ab with TSH in benign and malignant thyroid tissue homogenates were determined at four different temperatures.

4. Thermodynamic parameters of the standard state $(\Delta H^\circ, \Delta G^\circ, \Delta S^\circ)$, the transition state $(\Delta H^*, \Delta G^*, \Delta S^*)$ and activation energy (E_a) were determined for the binding of ^{125}I-anti TSH Ab with TSH in benign and malignant thyroid tissue homogenates.

5. The bound and unbound of ^{125}I-anti TSH Ab with TSH were isolated by gel filtration technique in thyroid tissue of patients with benign and malignant homogenate in order to estimate the cpm and the absorbance of the complex and free fractions. The data obtained revealed that these radioactivity (cpm) and absorbance in case of benign was higher than that of malignant patients.

6. Spectroscopic studies in the U.V. range (200-320 nm) were carried out for ^{125}I-anti TSH Ab/TSH (complex) and unbound ^{125}I-anti TSH Ab (free) fraction were obtained and the effect of pH, solvent, polarity and temperature were studied.

Contents

List of Figures

Chapter One

Chapter Two

Chapter Three

List of Tables

CHAPTER 1

Introduction

Chapter One

Introduction

1.1. Thyroid Gland Anatomy and Histology

he thyroid gland originated as an out pouching in the floor of the pharynx, which grows downward anterior to the trachea[1,2]. The fully developed thyroid gland in man is composed of two lobes connected by a thin isthmus, which gives the gland the appearance of a butterfly [1] (Figure 1.1).

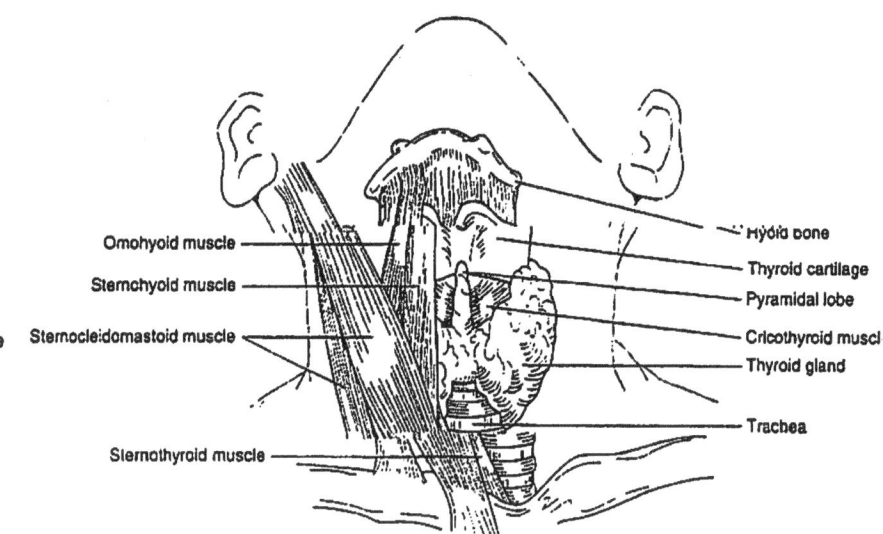

Omohyoid muscle
Sternchyoid muscle
Sternocleidomastoid muscle

Sternothyroid muscle

Hyoid bone
Thyroid cartilage
Pyramidal lobe
Cricothyroid muscl
Thyroid gland

Trachea

Figure (1.1): Gross anatomy of the human thyroid gland (anterior view)[1].

Each lobe is pear-shaped and measure about 2.5-4.0 cm in length, 1.5-2.0 cm in width, and 1.0-1.5 cm in thickness [3].

1

The weight of the gland in normal individuals varies depending on dietary ioidine intake, age and body weight, but in adults is approximately 10-20 g[4]. The thyroid gland has a rich blood supply with blood flow about 5 ml/g/min [1].

Transverse section of the neck at the level of thyroid isthmus shows the relationships of the thyroid gland to the trachea, esophagus, carotid artery and jugular vein (Figure 1.2).

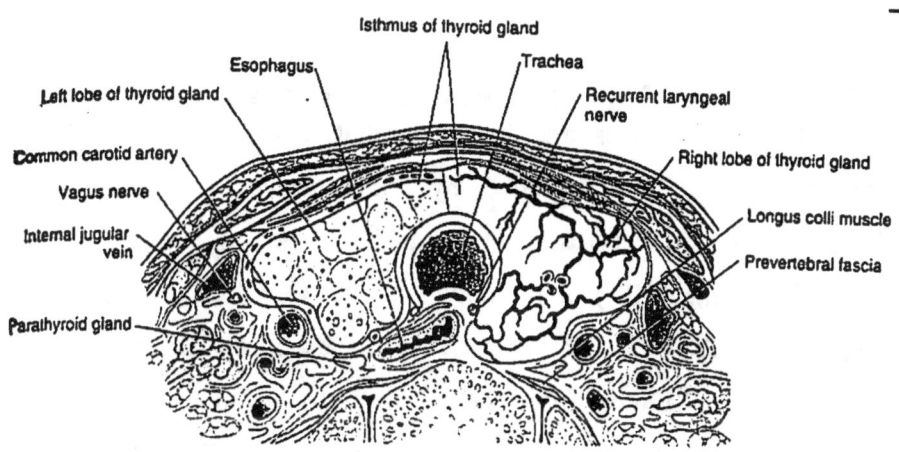

Figure (1.2): Cross section of the neck at the level of T1, showing thyroid relationship [1].

Other important anatomic consideration includes the two pairs of parathyroid glands that usually lie behind the upper and middle thyroid lobes [5].

On microscopical examination, the thyroid gland is found to consist of series of follicles, the secretory units of the gland, which contain an amorphous material called colloid and are surrounded by a single layer of thyroid epithelial cells. Colloid is mainly composed of thyroglobulin (an iodinated glycoprotein). Numerous microvilli project from the surface of the follicle into the lumen; these are involved in endocytosis of thyroglobulin, which is then hydrolyzed in the cell to release thyroid hormones [6]. The thyroid gland also contains another type of cell known as parafollicular or C cells. These cells have been shown to produce the polypeptide hormone calcitonin [1].

1.2. Thyroid Gland Physiology

The thyroid gland is the largest organ specialized for endocrine function in the human body [7]. Its function is to secrete a sufficient amount of thyroid hormones. Primarily, 3,5,3', 5'-L-tetraiodothyronine (also called thyroxine [L-T_4]), and a lesser quantity of 3,5,3'-L-triiodothyronine (L-T_3)[8]. Thyroid hormones promote normal growth and development and regulate a number of homeostatic function, including energy and heat production [6].

1.2.1. Thyroid hormones structure

Thyroid hormones are unique in that they contain 59-65% of trace element iodine [8]. The iodinated thyronines derived from iodination of the phenolic rings of tyrosine residues in the thyroglobulin to form monoiodotyrosine (MIT) or diiodotyrosine (DIT), which are coupled to form L-T_4 (DIT + DIT) or L-T_3 (MIT + DIT). In addition, the thyroid gland secretes small amounts of biologically inactive 3,3', 5'-L-triiodothyronine (also called reverse L-T_3 [rL-T_3]) and minute quantities of MIT and DIT [9]. The structure of Thyroid hormones and other derivatives are shown in Figure (1.3).

$$2I^- + H_2O_2 \rightarrow I_2$$

Figure (1.3): Structure of thyroid hormones and related compounds[9].

1.2.2. Thyroid hormones biochemistry

Dietary iodine (I^-) is essential for synthesis of Thyroid hormones [9]. The synthesis of L-T_4 and L-T_3 by the thyroid gland involves six major steps as shown in Figure (1.4).

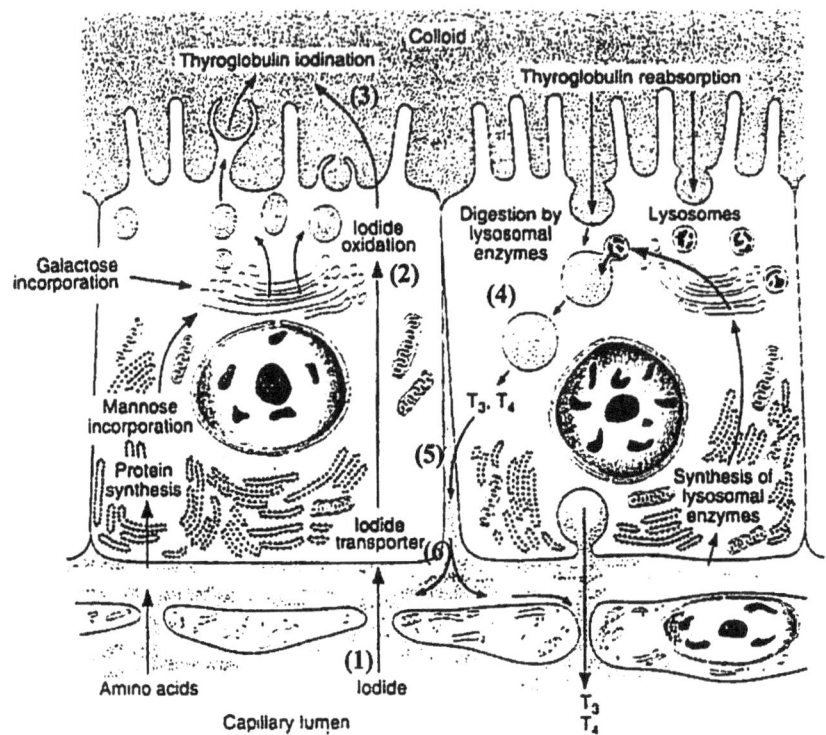

Figure (1.4): Thyroid hormones synthesis and release [5].

All these steps in addition to thyroid cell growth and development are stimulated by the pituitary hormone, thyroid stimulating hormone (TSH)[10], these steps are:

1. Active transport of I⁻ across the basement membrane into the thyroid cell (trapping of iodide);

2. Oxidation of iodide and iodination of tyrosyl residues in thyroglobulin;

3. Coupling of iodotyrosine molecules within thyroglobulin to form L-T$_3$ and L-T$_4$;

4. Proteolysis of thyroglobulin, with release of free iodothyronines and iodotyrosines;

5. Deiodination of iodotyrosines within the thyroid cell, with conservation and reuse of the librated iodide;

5

6. Under certain circumstances, intrathyroidal 5'-deiodination of L-T$_4$ to L-T$_3$ [10,11].

In blood circulation L-T$_4$ and L-T$_3$ are tightly bound to serum carrier proteins: thyroxine binding globulin (TBG), thyroxine binding prealbumin (TBPA) and albumin (TBA). The unbound or free fractions are the biologically active fractions and represented only 0.04% of the total L-T$_4$ and 0.4% of the total L-T$_3$ [12].

Approximately 40% of the secreted L-T$_4$ is deiodinated by the liver and other peripheral tissues to yield L-T$_3$, and about 45% is deiodinated to yield rL-T$_3$. Therefore, L-T$_4$ is considered as prohormone and L-T$_3$ is 4 to 5 times more potent in biological systems than L-T$_4$ [13].

Thyroid hormones enter cells, and L-T$_3$ binds to receptors in nuclei. L-T$_4$ can also bind, but not as avidly. There are two thyroid hormones receptors, hTR$_\alpha$ and hTR$_\beta$. The thyroid hormone-receptor complex then binds to DNA and increases the expression of specific genes. The resultant mRNAs trigger the production of various enzymes that alter cell function [14].

1.2.3. Thyroid hormones regulation

The growth and function of thyroid gland and the peripheral effect of thyroid hormones are controlled by at least four mechanisms [15]:

1. The hypothalamic-pituitary-thyroid axis (Figure 1.5), acting in a negative-feedback cycle, in which hypothalamic thyrotropin-releasing hormone (TRH) stimulates synthesis and release of anterior pituitary TSH, which inturn stimulates growth and hormone secretion by the thyroid gland;
2. The pituitary and peripheral deiodinases, which modify the effects of L-T$_4$ and L-T$_3$;
3. Autoregulation of hormone synthesis by the thyroid gland itself in relationship to its iodine supply;
4. Stimulation or inhibition of thyroid function by TSH receptor antibodies [15].

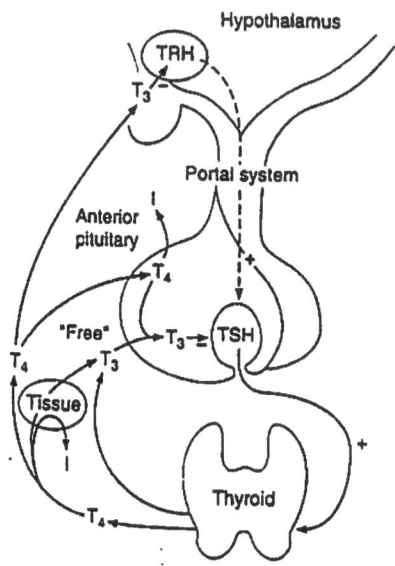

Figure (1.5): The hypothalamic-hypophysial-thyroid axis [13].

1.3. Pathological Conditions of Thyroid Gland
1.3.1. Hyperthyroidism

Clinical syndrome caused by excess of circulating active thyroid hormones [16]. Characterized by weight loss, sweating, fatigue, tachycardia, general muscle weakness arterial fibrillation and even angina [17]. It is either primary or secondary hyperthyroidism [18]:

1. Primary hyperthyroidism: In which the disorder is in thyroid gland itself.

    a. Toxic multinodular goiter.

    b. Thyroid adenoma.

    c. Thyroid carcinoma.

2. Secondary hyperthyroidism: In which the cause is in a site other than thyroid gland.

    a. Increased TSH.

    b. Grave's disease.

c. Neonatal hyperthyroidism (mother with thyroid stimulating immunoglobulin).

d. Pituitary tumors.

e. Exogenous adminstration of excessive thyroid hormones.

1.3.2. Hypothyroidism

Clinical syndrome caused by insufficient amount of thyroid hormones [8]. Characterized by lethargy, tiredness, cold intolerance, dryness and coarsening of skin and hair, weight gain, slow relaxation of muscles and tendon reflexes [16]. It is either due to primary or secondary causes [18]:

- **Primary Causes:**
  a. Atrophic hypothyroidism.
  b. Autoimmune hypothyroidism (Hashimoto's thyroiditis).
  c. Post-surgery, radioactive iodine, antithyroid drugs.
  d. Congenital (cretinism).
- **Secondary Causes:**
  a. Pituitary or hypothalamic disease.
  b. Iodine deficiency (endemic goiter).

1.4. Thyroid Tumors

Thyroid tumors are the most common endocrine neoplasms [19]. In 95% of cases, Thyroid tumor presents as a nodule or lump in the Thyroid [20]. In contrast to Thyroid nodules, thyroid carcinoma is a rare condition with an incidence of 0.004% per year according to the Third National Cancer Survey (2001). Thus, most Thyroid nodules are benign, and is important to identify those that are likely to be malignant [21].

1.4.1. Benign thyroid tumors

Benign tumors are one of the important causes of nodularity in thyroid gland. The adenoma, which is the most common type, varies in size and histologic characteristics and are often classified into three major types; papillary, follicular and Hurthle cell. The more highly differentiated adenoma (follicular) is the most common and is the most likely to mimic the function of normal thyroid tissue [22]. Though their function may be responsive to TSH stimulation, it differs from that of normal thyroid tissue in being autonomous, i.e., the basal activity is independent of TSH stimulation [23].

Initially, adenoma function is insufficient to disturb hormonal equilibrium. With time the nodule grows larger, its function increasing until it is sufficient to suppress TSH secretion [24]. At this time, the patient may or may not be overtly thyrotoxic, but frank thyrotoxicosis usually supervenes eventually (toxic adenoma) which revealed by radioiodine thyroid scan as a hot nodule [25].

1.4.2. Malignant thyroid tumors

Most thyroid carcinomas are characterized by an indolent course with low morbidity and mortality [26]. Although they account for less than 1% of all cancer deaths, they command the attention because they most often present as a thyroid nodule; therefore, they must be identified from the many other more common causes of thyroid nodules seen in 4% to 7% of the populations [27]. Furthermore, although the clinical course of most thyroid carcinoma is among the most prolonged of all human carcinomas, they do cause death in sufficient number to justify early diagnosis and treatment [28].

1.4.2.1. Incidence

Thyroid carcinoma is the most common type of endocrine malignancy and accounts for most of the deaths (about 63%) due to endocrine cancer [29]. In

9

general there is a 5% to 10% chance of malignancy in all thyroid nodules for total population [30].

The annual incidence of thyroid carcinoma is between 0.5 and 1.0 per 100000 populations in most countries, and the most recent global estimate suggests a total of 87000 new cases worldwide each year [31]. Clinical thyroid malignancy is relatively uncommon, accounting for about 2% of human malignancies [29]. Nonetheless, thyroid carcinomas in 2000 accounted for 91% of endocrine malignancies in the United States [32].

In most countries, incidence rates for papillary thyroid carcinoma generally exceed those for follicular thyroid carcinoma [33,34]. Very similar data on the frequency of various histotypes are contained in three large series of more than 97000 cases of thyroid carcinoma reported from both the United States and Japan [35,36].

Moreover, the incidence of thyroid carcinomas in term of sex is approximately two to four times more frequent in females compared with males [37]. On the other hand, the difference in the incidence in terms of race is that, the well-differentiated thyroid carcinoma of papillary subtype has a greater incidence, while, that of follicular subtype has a lower incidence in caucasians than in African-Americans of both sexes [38].

1.4.2.2. Etiology

External radiation to thyroid during childhood is the only well-established causative factor for thyroid carcinoma (mainly papillary) [39,40]. In Belarus and Ukraine, after the Chernobyl nuclear accident, the incidence of thyroid carcinoma started to increase as early as 4 years, suggests that radioactive isotopes have a direct tumorigenic effect on the thyroid [41,42].

Although a prolonged stimulation to thyroid galnd by TSH due to iodine deficiency and goitergenous drugs showed to produce thyroid tumors in experimental animals [43], there is no evidence of a primary TSH-related induction of thyroid tumors in humans [44]. However, the incidence of papillary carcinomas being more frequent in areas where dietary iodine is high [45].

A familiar occurrence of papillary carcinoma has been reported 3% of cases [46]. Other factors such as, diet, body weight and number of pregnancies may modify the risk for thyroid carcinomas [47].

1.4.2.3. Histological classification

There are many types of thyroid tumors, each with a distinctive epidemiology, natural history, treatment and prognosis [19]. In general, malignant thyroid tumors can be divided, according to cell of origin, into three main groups [44]:

1. Tumors arising from the thyroid cells. That is, of follicular cell origin. The WHO classification subdivides malignant tumors of follicular cell origin into undifferentiated (anaplastic) carcinoma and differentiated thyroid carcinoma, either papillary of follicular carcinoma [48].

2. Medullary carcinoma arising from the parafollicular cells (called also C-cells).

3. Primary thyroid lymphoma and sarcoma arising from immune and stromal cells respectively, (Table 1-1).

Table (1.1): Classifications of thyroid neoplasms [23].

Primary epithelial tumors	Tumors of C cells
Tumors of follicular cells	Medullary carcinoma
Benign: follicular adenoma	Tumors of follicular and C cells
Malignant: carcinoma	Mixed medullary-follicular carcinoma
Differentiated	Primary nonepithelial tumors
Papillary	Malignant lymphomas
Follicular	Sarcomas
Poorly differentiated	Others
Insular	Secondary tumors
Others	
Undifferentiated (anaplastic)	

- **Papillary Carcinoma**

Papillary carcinoma comprises approximately 80% to 85% of malignant thyroid tumors in developed countries [49,50]. Most papillary carcinomas are intrathyroidal and demonstrate a partial encapsulated or diffusely infiltrated borders. The tumor is usually a firm white mass that measures typically between 2 and 3 cm [51] and it is multicentirc in 20% to 80% of patients and bilateral in about one third [52].

Papillary thyroid carcinoma is characterized by typical nuclear features include large-sized and overlaping nuclei with invagination of cytoplasm into nuclei [53]. Psammoma bodies, present in 40% to 50% of cases and are almost pathognomonic of this carcinoma [54].

- **Follicular Carcinoma**

Follicular carcinoma is characterized by follicular differentiation but without the nuclear changes characteristic of papillary carcinoma [55]. Two forms are recognized according to the pattern of invasion: minimally invasive and widely invasive follicular carcinomas. They are encapsulated and invasion of the capsule and vessels are the key feature distinguishing follicular carcinomas from adenomas [26].

- **Anaplastic Carcinoma**

Anaplastic carcinoma of the thyroid is one of the most aggressive carcinomas encountered in humans [56]. In most cases, it represents the terminal stage of the dedifferentiation of a follicular or papillary carcinomas [57]. In fact, anaplastic cells do not produce thyroglobulin, they are not able to transport iodine, and thyrotropin receptors are not found in their plasma cell membrane[56].

The tumor is typically composed of varying proportions of spindle, polygonal and gaint cells. Keratin is the most useful epithelial marker and is present in 40% to 100% of tumors [58].

• **Medullary Carcinoma**

Medullary thyroid carcinoma is a tumor of the C-cells [59]. Grossly, it presents as a white-red, hard lesion [60]. At microscopy, it presents as sheets of spindle or round cells, typical of neuroendocrine tumors. Nuclei are usually uniform with rare mitosis. Deposits of amyloid are frequently found [58].

1.4.2.4. Staging

While there are many staging systems for thyroid carcinomas, the tumor, node, metastasis (TNM) system, which provided in 1997, is the most widely used [61,62] (Table 1.2). In general, thyroid carcinoma staging is unique in that both the histologic diagnosis and the age of the patient are included because of their prognostic importance [37].

Table (1.2): Staging of thyroid gland carcinomas [26].

Stage	Papillary of follicular		Medullary	Descriptor
	< 45 years	> 45 years		
I	Any T, Any N, MO	T1, NO, MO	T1, NO, MO	T1: ≤ 1 cm in gretaest dimention; limited to thyroid NO: No regional lymph node metastasis MO: No distant metastasis
II	Any T, Any N, M1	T2-3, NO, MO	T2-4, NO, MO	T2: Tumor > 1 cm but < 4 cm in greatest dimension; limited to thyroid T3: Tumor > 4 cm in greatest dimension; limited to thyroid. T4: Tumor any size extending beyond thyroid capsule M1: Distant metastasis
III		T4, NO, MO Any T, N1, MO	Any T, N1, MO	N1 Regional lymph node metastasis°
IV		Any T, Any N, M1	Any T, Any N, M1	

On the basis of TNM staging system, all patients younger than 45 years of stage with papillary (PTC) and follicular (FTC) carcinomas have stage I disease unless they have distant metastases, in which the disease is classified stage II.

Older patients (age 45 or older) with node-negative papillary or follicular carcinomas (T_1NoMo) have stage I disease. Intrathyroidal tumors 1.1 cm or larger are stage II, and either nodal involvement or extrathyroid invasion in older patients with PTC or FTC leads to stage III classification [61-63].

For medullary thyroid carcinoma (MTC), the scheme is similar in that tumor 1 cm or smaller is stage (I) and node-positive is stage (III). There is no age distinction for MTC; however, the local (extrathyroid) invasion is defined as stage II. For both MTC and older patients with PTC or FTC, stage IV denotes the presence of distant metastases. Independent of age or tumor extent, all patients with undifferentiated anaplastic thyroid carcinoma (ATC) are considered to have stage IV disease [61-63].

1.4.2.5. Spread

The pathway and degree of spread of thyroid cancer are different in various subtypes. For example, papillary carcinoma spread through the lymphatic within the thyroid to the regional lymph nodes and, less, frequently, to the lungs [64]. While in follicular carcinoma, the multicentricity and lymph node involvement are less frequent than papillary carcinoma [65].

On the other hand, both metastatic spread to regional lymph nodes, which occurs early and distant metastases, which usually slow growing, are frequent in medullary thyroid carcinoma. The sites most frequently affected are the liver, lungs and bones [66]. Moreover, anaplastic carcinoma has a rapidly progressing course. At time of diagnosis, many patients have metastases in lymph nodes and invasion of adjacent organs. Distant metastases ultimately occur in many patients, most commonly in the lungs, followed by bones liver and brain [67].

1.4.2.6. Clinical features (symptoms and signs)

The thyroid carcinoma may present itself in one or more of the following symptoms and signs:

- Painless lump in neck, which is either due to growth within the thyroid gland (nodule) or metastasis to regional lymph nodes [68]. Some signs that the nodule may be cancerous include; **(1)** single and hard nodule that is grows fast; **(2)** solid mass in ultrasonography; **(3)** cold nodule in thyroid scan [69].

- Signs of pressure or invasion of surrounding tissues are present in anaplastic and long standing tumors. These signs include hourseness of voice, dysphagia and suffocation [70].

- Metastatic functioning differentiated thyroid carcinoma can sometimes secrete enough thyroid hormone to produce thyrotoxicosis [71].

- Medullary carcinoma frequently causes flushing and diarrhea (30%), fatigue and about 5% developed Cushing's syndrome from secretion of ACTH or CRH [72].

1.4.2.7. Diagnosis

In addition to a complete medical history and physical examination, diagnostic procedures for thyroid carcinoma inculde:

- **Serum Factors**

A high titre of thyroid autoantibodies in serum suggests chronic thyroiditis but does not rule out on associated malignancy [73]. Howevere, an elevated serum calcitonin, particularly in patients with a family history of medullary carcinoma, strongly suggests presence of thyroid carcinoma [74]. Elevated serum thyroglobulin following total thyroidectomy for papillary or follicular thyroid carcinoma usualy indicates metastatic disease [75].

- **Imaging Studies**

    Scanning procedures can be used to identify "hot" or "cold" nodules [76]. Hot nodules are almost never malignant, whereas cold ones may be [77]. Thyroid ultrasound can distinguish cystic from solid lesions [76]. Pure cystic is almost never malignant [78]. CT scanning or MRI may be helpful in defining substential extension or deep thyroid nodules in the neck [79].

- **Fine Needle Biopsy**

    The major advances in management of the thyroid nodule in recent years has been the fine needle aspiration biopsy, in which a sample of the nodule taking with a needle for examination under a microscopic [80]. The technique is diagnostic in about 95% of all types of thyroid malignancies except for follicular carcinoma which can not distinguished by cytology [81], thus it is always suspicious of malignancy.

1.4.2.8. Tumor markers

The most useful analytes of thyroid carcinoma include; thyroglobulin (Tg), immunoreactive calcitonin (ICT) and carcinoembryonic antigen (CEA) [26].

- **Serum Thyroglobulin**

    Although serum Tg measurements are not useful in the preoperative distinction of follicular cell derived carcinomous (FCDC) from benign thyroid diseases, they have a useful role in the postoperative follow-up of such patients[82]. After the elimination of the thyroid tissue, increasing Tg levels may be useful indicators of the presence of metastatic FCDC, whereas subnormal or undetectable Tg levels indicate the absence of metastatic involvement [83].

    Two problems limited the use of serum Tg in postoperative management; the first problem is that most Tg assays are invalidated by the presence of Tg autoantibodies (AAb). Such AAb if present in sufficient levels will render the measured Tg levels falsely high or low [84]. The second problem relates to the fact that serum Tg assays have not been well standardized. Many clinicians use Tg levels above 5 ng/ml to indicate the need for imaging, in search of possible persistent or recurrent disease [85].

- **Immunoreactive Calcitonin (ICT)**

    Serum ICT is the classic tumor marker used in both diagnosis and follow-up of patients with meduallry thyroid carcinoma (MTC) [86]. In general, the basal and stimulated ICT levels are though to be somewhat proportional to the MCT tumor mass [87]. It is also true that many patients with MTC who undergo surgical treatment may still have postoperative elevation in ICT levels without obvious clinical or imaging evidence of persistent disease; furthermore, slowly increasing ICT levels may not necessarily imply a substantial worsening of prognosis [88].

- **Carcinoembryonic Antigen (CEA)**

    A second major tumor marker for MCT, which is often overlooked, is the readily measured CEA [89]. In general, the CEA level is higher in more malignant MTC, hence some authorities suggest that an increasing CEA level postoperatively is more likely to correlate with aggressive recurrent tumor [90].

1.4.2.9. Treatment

The initial treatment of thyroid carcinoma is surgical removal of tumor mass with the need for extensive regional lymph nodal dissection, the so-called primary treatment. Subsequently, ablative therapy is introduced as almost all patients with FCDC required thyroid hormone therapy and radiotherapy in postoperative period [91]. In addition, anticancer chemotherapy required in special cases [92].

The main principles in treatment of thyroid carcinomas can be summarized as follows [93]:

- **Surgery**

  Used as initial therapy for all thyroid cancer subtypes. It includes different operations such as lobectomy and subtotal, total or near total thyroidectomy.

- **Thyroid Hormone Therapy**

  Used as ablative therapy in papillary carcinoma after surgery to suppress the pituitary gland from secreting more TSH, which may stimulate a recurrence of this cancer [94].

- **Radiotherapy**

  o *Remnant radioactive iodine ablation (RRA)*

  RRA is the second most frequently used postoperative adjuvent therapy for patients with FCDC and it has been defined as the destruction of residual macroscopically normal thyroid tissue after surgical thyroidectomy [95].

  o *Radioactive iodine (RAI) Therapy*

  In which larger adminstrated doses of I^{131} are used in an attempt to destroy persistent neck disease or distant metastatic lesions [96].

- **External Irradiation**

  External irradiation is rarely used as adjuvant therapy in the initial management of patients with FCDC or MTC [97]. On the other hand, it is routinely used in treatment of primary thyroid lymphoma and postoperatively for anaplastic thyroid carcinoma [98].

- **Chemotherapy**

  In which a single or combination of anticarcinoma cytotoxic drug is used. In patients with differentiated FCDC, chemotherapy restricted to those tumors that are surgically unresectable or those unresponsive to RAI or external irradiation [99]. Also it has rarely been used in the primary management of MTC[100]. In contrast, in disseminated thyroid lymphoma, the chemotherapy is treatment of choice [98].

1.4.3. Thyroid stimulating hormone (TSH)

Thyrotropin (thyroid stimulating hormone, TSH) is one of the carbohydrate-containing polypeptide hormones and it has considerable immunological and chemical similarity to three other human glycopeptide hormones, luteinizing hormone (LH), follicle stimulating hormone (FSH) and chornionic gonadotropin (hCG) [101]. TSH plays a key role in regulation of thyroid gland function [102].

1.4.3.1. Chemistry and mechanism of TSH

Human TSH is a glycoprotein (Mw 28000 Da) secreted by the tropic cells of the anterior pituitary gland that contains 211 amino acid residues, plus hexoses, hexosamines and sialic acid [101]. It is made up of two non-covalently linked α and β subunits. The structure of α-subunit of TSH is identical to that of other glycoprotein hormones (FSH, LH and hCG) but the β-subunit differs in these glycoproteins and is responsible for their biologic and immunologic specificity[103].

Glycosylation of both subunits begins with attachment in the rough endoplasmic reticulum of a 14-sugar unit to asparagine residues with modification of these carbohydrate side chains in the Golgi apparatus. The initial glycosylation is necessary for proper combination of the α and β subunits and the final pattern of carbohydrate side chains is necessary for full biologic activity, probably because it produce the proper folding of the molecule. In addition, deglycosylated TSH is removed more rapidly from the circulation; thus its effectiveness is decreased [104].

The biologic half life of human TSH is about 60 minutes. TSH is degraded for the most part in the kidneys and to a lesser extent in the liver [105].

19

Secretion is pulsatile and mean output starts to rise at about 9:00 P.M., peaks at midnight and then declined during the day. The normal secretion rate is about 110 μg/dL and the average plasma level is about 2 μU/mL [106].

1.4.3.2. Regulation of TSH secretion

The secretion of TSH is controlled by both stimulatory (TRH) and inhibitory (somatostatin) influences from the hypothalamus and in addition is modulated by the feedback inhibition of thyroid hormone on hypothalamic-pituitary axis [107].

The response of TSH to TRH is modulated by the circulating levels of thyroid hormones [108]. Administration of TRH increases TSH within 2 minutes, and this response is blocked by previous L-T$_3$ administration; however, larger dose of TRH may overcome this blockade suggesting that both L-T$_3$ and TRH act at the pituitary level of influence TSH secretion. On the other hand, somatostatin, which is an inhibitory hypothalamic peptide, augments the direct inhibitory effect thyroid hormones on the thyrotrophs and hence inhibits TSH release from anterior pituitary gland [1].

In addition to these hypothalamic influences on TSH secretion, other neurally mediated and endocrinal factors may be important. Dopamine physiologically inhibits TSH secretion as well as blunts the TSH response to TRH [109]. Moreover, glucocorticoid excess has been shown to impair the sensitivity of pituitary to TRH and may lower TSH to undetectable levels. However, estrogens increase the sensitivity of the thyrotroph to TRH and women have a greater TSH response to TRH than men do [110] (Figure 1.6).

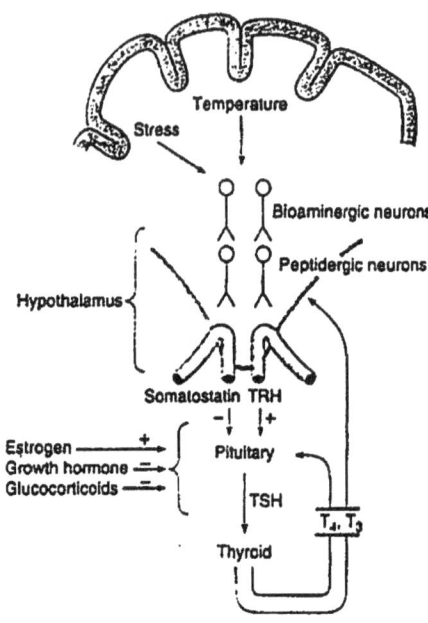

Figure (1.6): Regulation of TSH secretion [1].

1.4.3.3. Action of TSH on thyroid cells

TSH has many actions on thyroid cells. Most of its action is mediated through the G-protein-adenylyl cyclase-cAMP system (Figure 1.7).

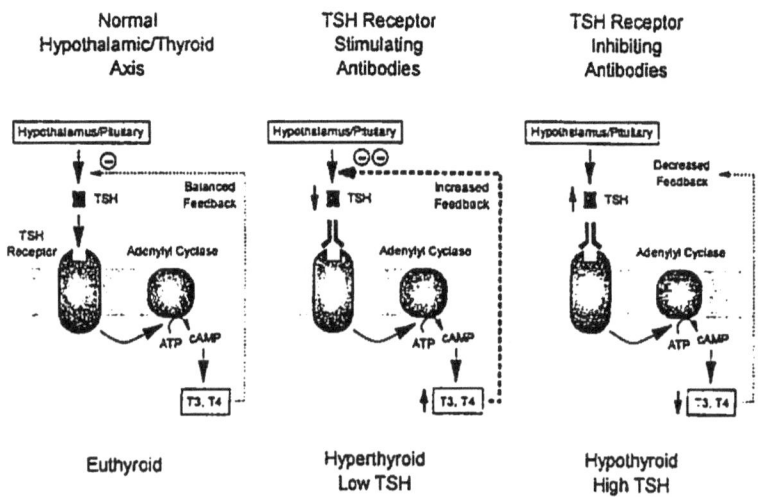

Figure (1.7): Mechanism of actions of hormones that act at the cell surface[23].

The activation of the phosphotidylinositol (PIP$_2$) system with increase in intracellular calcium may also be involved [111]. The major actions of TSH include the following [112]:

- **Changes in Thyroid Cell Morphology**

TSH rapidly induce pseudopods at the cell colloid border, accelerating thyroglobulin resorption, colloid content is diminished. Intracellular colloid droplets are formed and lysosome function is stimulated, increasing thyroglobulin hydrolysis.

- **Cell Growth**

Individual thyroid cells increase in size, vascularity is increased, and, over a period of time, thyroid enlargement or goiter develops.

- **Iodine Metabolism**

TSH stimulates all phases of iodide metabolism, from increased iodide uptake and transport to increased iodination of thyroglobulin and increased secretion of thyroid hormones.

- **Others**

TSH has many other effects on the thyroid gland, including stimulation of glucose uptake, oxygen consumption, CO_2 production and an increase in glucose oxidation via the hexose monophosphate pathway and Kreb's cycle. This is accelerated turnover of phospholipids and stimulation of synthesis of purine and pyrimidine precursors, with increased synthesis of DNA and RNA [112].

1.4.3.4. TSH receptors

TSH achieves its effect on thyroid gland by binding to a specific TSH receptor (TSH-R) on thyroid cell membrane and activating both the G-protein-adenylyl cyclase-cAMP and the phospholipase C signaling systems [113]. The human TSH-R gene is located on chromosome 14q3 [114]. The TSH-R is a single chain glycoprotein containing 764 amino acids and like TRH receptor of anterior pituitary, the TSH-R in thyroid follicular cell is a member of the seven-membrane spanning, GTP-binding, protein-coupled receptor family [113].

Structurally, TSH-R can be divided into two subunits, subunit A, containing 397 amino acids, representing the ectodomain which is involved in ligand-binding; and subunit B, which includes the intramembrane and intracellular portion of the receptor involved in activation of thyroid cell growth, thyroid hormone synthesis and release (Figure 1.8) [115].

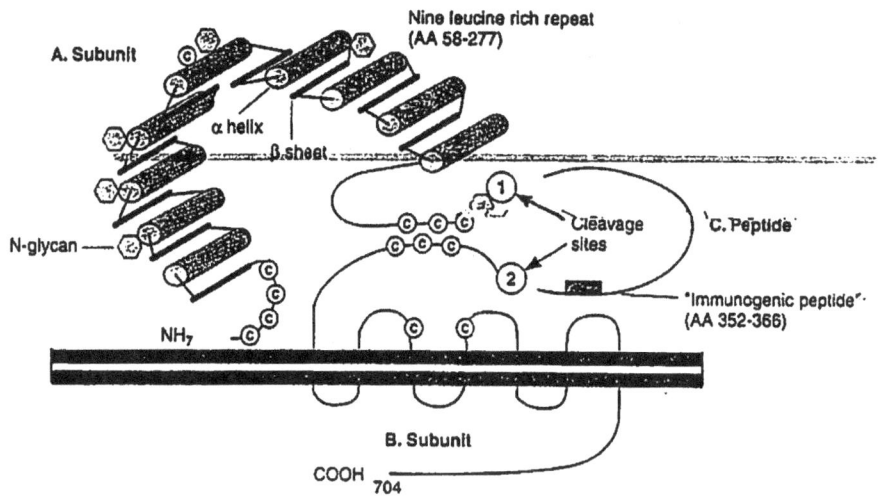

Figure (1.8): Schematic representation of the TSH receptor [115].

The A subunit is the ligand-binding portion of the receptor and the B subunit is the activation portion. The ligand which binds to the receptor includes TSH, TSH-stimulating antibody, and TSH-blocking antibody. There are two cleavage sites which allow breakage of the receptor and loss of the A subunit in the serum (C: cystine residue) [115].

Mutations in the TSH-R have been associated with either spontaneous activation of the receptor and clinical hyperthyroidism or with resistance to

TSH [116]. The TSH-R has been cloned [117], and specific mutations have been identified in association with hyperfunctioning follicular thyroid neoplasms[118,119], (Figure 1.9).

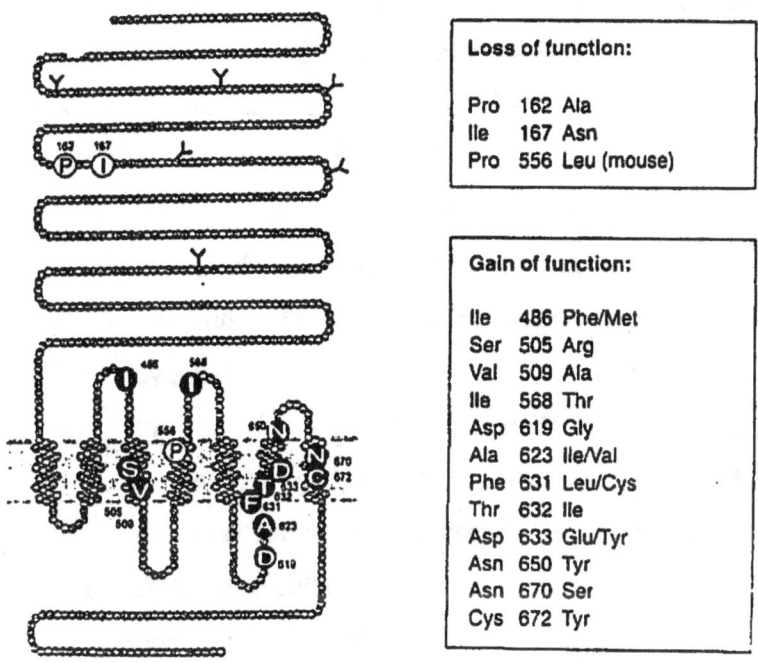

Loss of function:

Pro 162 Ala
Ile 167 Asn
Pro 556 Leu (mouse)

Gain of function:

Ile 486 Phe/Met
Ser 505 Arg
Val 509 Ala
Ile 568 Thr
Asp 619 Gly
Ala 623 Ile/Val
Phe 631 Leu/Cys
Thr 632 Ile
Asp 633 Glu/Tyr
Asn 650 Tyr
Asn 670 Ser
Cys 672 Tyr

Figure (1.9): Schematic diagram showing the mutations that lead to gain or loss of function of TSH receptor [23].

1.4.3.5. TSH-determination

Measurement of TSH was originally based on bioassays, such as stimulation of colloid droplet formation in the guinea pig thyroid gland and the release of labeled thyriodal iodide into mouse blood [120]. These early in vivo bioassays, however, were of limited sensitivity and precision and were not applicable to the measurement of TSH in unfractionated serum. Most TSH bioassays have involved the in vitro stimulation of thyroid cyclic adenosine monophosphate (cAMP) or adenylate cyclase activity [121]. Unfortunately such methods require purification and concentration of TSH from serum before assay. Sensitive detection of TSH in unfractionated serum is possible by using a cytochemical assay [122], but this procedure is technically difficult and time consuming.

At present, immunoassay is the standard procedure for the measurement of serum TSH in the clinical laboratory. The competition radioimmunoassay (RIA) and immunoradiometric assay (IRMA) are the most widely used immunoassay [123]. The determination of TSH by RIA proved very valuable in assessing the elevated TSH values in primary hypothyroidism. However, because it could only detect 1 mU of TSH per liter, it could not distinguish normal values from subnormal values associated with hyper thyrodism [124]. In contrast to the inverse dose-response curve found in RIA, immunometric assays have a positive dose-response curve, with higher levels of signal corresponding to higher concentration of TSH which offer not only improved sensitivity of TSH measurement, but also rapid turnaround time and a wider linear measurement range compared with competitive RIA [125]."In addition, other immunometric assays for TSH are available in which the detection antibody is labeled with various signal molecules, such as enzyme [126], fluorophore [127], or luminescent molecule [128]. Those assays use chemiluminescent technologies have enhanced precision at even lower detection limits (less than 0.02 mU/L) than IRMA [129].

1.4.3.6. TSH and thyroid carcinoma

The TSH is the principle hormone regulating the growth and function of thyroid gland and thus, excess TSH may be of etiologic importance in the development of thyroid carcinoma [130]. This hypothesis is supported by the observation that growth of some thyroid carcinoma depends on TSH secretion, so that suppression of TSH release by administration of thyroxin is often an effective treatment for thyroid carcinoma [22,25]. Experimental studies provide further support to this hypothesis. Sustained elevation of TSH induces thyroid tumors in rodents [131].

The cells of most differentiated thyroid tumors possess TSH receptors. Their number varies with the histological type, the highest being found in follicular well-differentiated tumors and the lowest in less differentiated ones [132]. Moreover, response to TSH stimulation, as assessed by adenylcyclase levels and radioiodine uptake, varies for a given number of TSH receptors [133].

TSH stimulation increases radioiodine uptake by all thyroid tissues able to pick-up ^{131}I [134], and also increased Tg release into the blood even by tumors unable to concentrate ^{131}I [135]. This shows that all tumors are TSH dependent.

TSH plays a major role in the control of thyroid cell proliferation, as shown in the in vitro culture of normal human thyroid cells [136]. Other factors, such as thyroid-stimulating immunoglobulines, may enhance thyroid cell proliferation [137].

Aim of the Work

The aim of the present work is outlined, according to the following points:

1. Determination of L-T$_3$ and L-T$_4$ and total TSH levels in sera of normal subjects and patients with thyroid diseases.

2. Determination of the optimum conditions of the binding of tTSH with ^{125}I-anti tTSH Abs in benign and malignant thyroid tissues such as those of binding capacity and the effect of various factors like (antigen concentration, antibody concentration, temperature, pH, time, salt and halides).

3. Determination of the kinetic and thermodynamic parameters of the binding reactions in benign and malignant thyroid tissues of patients with thyroid disease.

4. Isolation of (TSH/^{125}I-anti tTSH Ab) complex from unbound ^{125}I-anti tTSH Ab in benign and malignant thyroid homogenate tissues.

5. Spectroscopic studies on the complex (^{125}I-anti tTSH Ab/TSH) and unbound ^{125}I-anti tTSH Ab) in malignant thyroid homogenate tissues.

CHAPTER 2

Experimental Work

Chapter Two

Experimental Work

2.1. Chemicals, Instruments and Samples

2.1.1. Chemicals

ll common laboratory chemicals and reagents were of analar grade and were used without further purification. Tris[hydroxymethyl] amino-methane, $MgCl_2$, $MnCl_2$, NaCl, NaBr were obtained from Fluka Company-Switzerland.

Sucrose, Na, K-tartarate, $CuSO_4.5H_2O$, glycerol, Na_2CO_3, NaF, $ZnCl_2$, NaI, $CaCl_2$, $NaHCO_3$, Bovine serum albumin (BSA), acetic acid, foiln cio colteau, LiCl, NH_4Cl and KCl were obtained from BDH.

Gel sephadex-G150, blue dextran 2000 were obtained from Pharmacia Fine Chemicals-Sweden.

The immunoradiometric assay (IRMA) Kit for determination of thyroid stimulating hormone (TSH) and the radioimmunoassay (RIA) Kit for determination of thyroxine ($L-T_4$) were purchased from (Immunotech, a Beckman Coulter Company) while the RIA Kit for determination of triiodothyronine ($L-T_3$) was obtained from (DiaSorine, Stillwater, Minnesota, USA).

2.1.2. Instruments

The instruments used in this work were, LKB gamma counter type 1270 Rack gamma II, cooling centrifuge type Hettich, Shimadzu U.V-Visible recorder

28

spectrophotometer type U.V.-160, Pye-Unicam pH meter, Memmert water bath, SM-Shaker and Memmert incubator.

2.1.3. Patients

Three groups of thyroid disease patients were included in this study. Group I contained 29 patients with euthyroid multinodular goiter. Group II consisted of 14 patients with toxic goiter, of this group, nine patients (64.3%) suffering from Grave's disease and five patients (35.7%) suffering from toxic multinodular goiter. Group III consisted of 10 patients with thyroid carcinoma, of this group, seven patients (70%) suffering from papillary carcinoma, two patients suffering from follicular carcinoma and only one patient (10%) suffering from anaplastic carcinoma. In addition to ten healthy volunteers included in the study as a control group (group IV).

All patients were admitted for management to several general hospitals (including Baghdad, Al-Kindy and Saddam Teaching Hospital) and from many private hospitals (including Dar Al-Najaat, Janin and Al-Jadriah Private Hospitals). The histopathological findings of thyroid biopsies were under the supervision of specialists Dr. Hanoun Al-Nousairy and Dr. Newal Alash.

The patients were newly diagnosed and not underwent any type of therapy and those who suffered from any disease, other than thyroid disease, were excluded from this study.

The host information of all patients and control subjects were summarized in Table (2.1).

Table (2.1): The host information of the thyroid disease patients and normal controls.

Groups	Group Name	No. of total studied sbjects	Cases	No % of cases	Age (year)
I	Ethyroid multinodular goiter	29	-	-	10-55(32.5)
II	Toxic goiter	14	Grave's disease (diffuse goiter)	9(64.3)	14-*58(36)
			Toxic multinodular goiter	5(35.7)	22-32(27)
III	Thyoid carcinoma	10	Papillary carcinoma	7(70)	8-38(23)
			Follicular carcinoma	2(20)	40-46(43)
			Anaplastic	1(10)	64
IV	Normal controls	10	-	-	10-60(35)

2.1.4. Preparation of blood samples

Blood samples (5 ml) were obtained from patients undergoing by venipuncture just before operation. The collected blood was left for 20 min at room temperature. After coagulation, sera were separated by centrifugation at 2000 xg for 10 min, and kept at -20°C until assaying.

2.1.5. Collection of specimens

The tumor tissues were removed from thyroid tumor patients. The specimens were cut-off and immediately rinsed with ice-cold isotonic saline solution. They were collected individually in plastic receptacle and stored at -20°C until homogenization.

2.1.6. Statistical analysis

Student's t-test was used to determine if the mean values of studied parameters were significantly different in between the individual groups included in this work. P values < 0.05 were considered significant [138].

2.1.7. Preparation of thyroid tumors tissue homogenate

The frozen tissue was thawed, weighed, sliced finely with a scalped in Petri dish standing on ice bath. The slices were further minced with scissors then homogenized at 4°C in buffer solution (0.01 M) with a ratio of 1:3 (weight: volume) using a manual homogenizer. The homogenate was filtered through several layers of nylon gauze, and then centrifuged at 4000 xg for 15 min in a cooling centrifuge 4°C. The supernatant was used through the study.

2.1.8. Solutions

The sucrose-Tris buffer solution (0.01 M, pH = 7.4) was prepared by dissolving (0.606 g) of Tris(hydroxy methyl)aminomethane and (51.345 g) of sucrose in 450 ml distilled water. The required pH was adjusted by adding HCl solution, and then the volume was completed to 500 ml by distilled water [94].

2.1.9. Preparation of hitsopathology slides

The thyroid tissue specimens were fixed in 10% formalin. Then the specimens processing in alcohol and clearing in xylol, and later embaded in parafin. Sections (5μm thickness) were taking by microtome instrument and put on glass slide. The slides were then stained by haematoxylin eosin stain [150].

2.2. Determination of Total Protein Content in Benign and Malignant Thyroid Tissue Homogenates

The total protein of thyroid tissue was determined by Lowry et. al., [139] method by using bovine serum albumin (BSA) as the standard protein.

- Solutions

  1. **Reagent A:** Alkaline sodium carbonates solution (2% Na_2CO_3 in 0.1 N NaOH).

2. **Reagent B:** Copper sulfate-sodium potassium tartarate solution (0.5% $CuSO_4.5H_2O$ in 1%, Na, K tartarate). This solution was prepared freshly by dissolving 0.1 g of Na, K tartarate in 10 ml of $CuSO_4.5H_2O$.

3. **Reagent C:** Alkaline copper solution. Mix 50 ml of reagent A with 1 ml of reagent B. Discard after one day.

4. **Reagent D:** Folin cio calteau, reagent prepared by the dilution of the commercial reagent with an equal volume of distilled water on the day of the use.

5. **Standard bovine serum albumin:** (BSA 0.2 mg/ml), working BSA solutions were prepared by serial dilutions of stock solution.

- **Method**

1. One milliliter of each of standard BSA (0, 20, 40, 80, 120 and 160) µg/ml was pipetted in a set of test tubes. The expriment was carried out in duplicate.

2. Twenty microliter of thyroid tumor homogenate was also pipetted in test tubes and the volume was made up to one milliliter with distilled water.

3. Five milliliter of reagent C was added to all assays tubes. Then the contents were mixed by vortexing and allowed to stand for 10 min at room temperature.

4. One half millileter of reagent D was added drop by drop with vigorous mixing to all assay tubes. The mixture was left to stand for 30 min. at room temperature.

5. The absorbance of the developing colour was read at 600 nm against the appropriate blank.

- **Calculations:**

The standard curve was obtained by plotting the absorbance against the corresponding concentrations of standard protein and used to determine the unknown protein concentration of the sample (thyroid tumor homogenate) as shown in Figure (2.1).

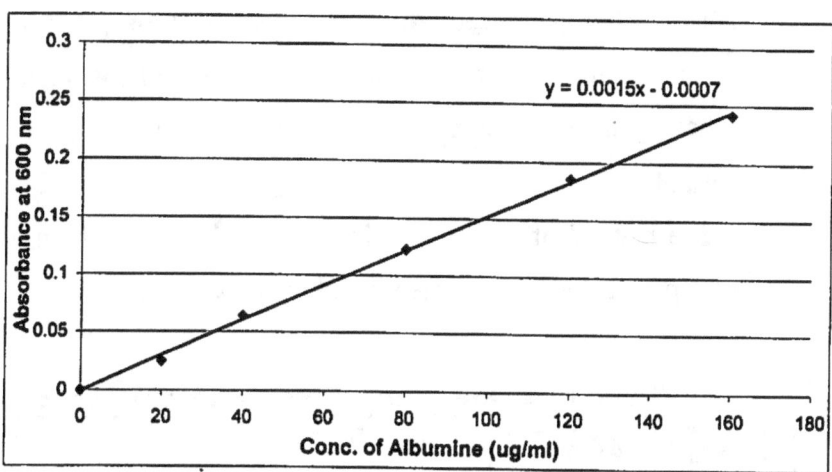

Figure (2.1): Standard curve of protein determination by Lowry's method.

2.3. Determination of TSH Levels in Sera of Benign and Malignant Thyroid Disease Patients and Controls

Serum levels of total TSH were determined by an Immunoradiometric method. The sample or standard were incubated in coated tubes with the first monoclonal antibody (Ab) in the presence of the second monoclonal ^{125}I-labled antibodies. After incubation, the contents of the tubes were decanted and washed, the unbound Ab was removed and the bound radioactivity was measured in a gamma counter [140].

- **Reagents:**

  1. Monoclonal ^{125}I-labeled anti-TSH antibody; 1 vial (11 ml), ready to use. The vial contains less than 296 kBq of ^{125}I-labeled antibody in buffer with proteins, sodium azide (< 0.1 %) and a dye.

  2. Anti-TSH monoclonal antiboy-coated tubes; 100 tubes; ready for use.

  3. Standards; 6 vials (1 ml) and 1 vial of "zero" standard, ready for use. The standard vials contain concentration of TSH for standard range from 0 to 50 IU/mL in buffer with sodium azide (< 0.1%).

  4. Control sera; 2 vials (1 ml); lyophilized.

5. The solution concentration was washed (20x); 1 vial (50 ml). The concentration must be diluted by 950 ml of distilled water and homogenized before use.

- **Method:**

The method of TSH determination was based on the use of TSH-IRMA kit assay protocol as summarized in Table (2.2).

Table (2.2): Summary of the assay protocol of TSH.

Step 1 (Addition)	Step 2 (Incubation)	Step 3 (Washing and Counting)
Adding successively to Ab-coated tubes • 200 µL of standard control or sample and • 100 µL of tracer • Vortex	• Incubation for 2 hrs (at 18-25°C) With • Shaking (at least 350 rpm)	• Decant The contents of tubes carefully (except the two tubes) • Rinsing twice the tubes with 2 ml of wash solution • Count bound cpm (B). • Total cpm (T).

- **Calculations:**

1. The mean count for each group of tubes was counted in a gamma-counter for 1 min.

2. The B/T ratio was computed for each standard and unknown samples as follows:

$$B/T\% = \frac{\text{Standard or sample mean counts}}{\text{Total activity mean counts}} \times 100$$

3. The standard curve was drowning by plotting the percent value of each standard against the corresponding standard concentration.

4. Locate c.p.m bound (or the ratio B/T) on the vertical axis of the standard curve read off the TSH concentration of the sample on the horizontal axis in IU/ml (Figure 2.2).

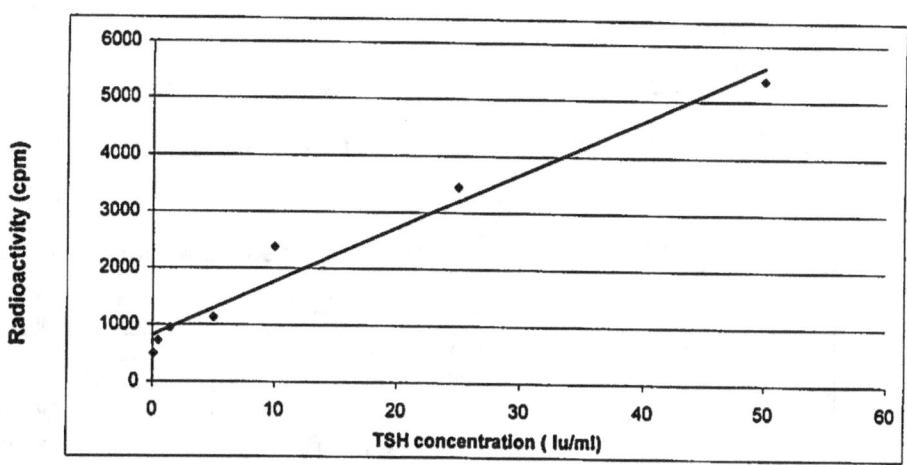

Figure (2.2): Standard curve of total TSH in human serum samples.

2.4. Determination of Total Thyroxine (L-T$_4$) levels in Sera of Benign and Malignant Thyroid Tumor Patients and Controls

Unknown samples and standards were incubated together with ^{125}I-thyroxine in monoclonal anti-T$_4$ antibody coated tubes. After incubation the contents of the tubes were decanted and the bound activity was measured in a gamma counter. The concentration of L-T$_4$ was reversely proportionated to the radioactivity measured [141].

- **Reagents:**

  1. ^{125}I-labled thyroxine; 1 vial (55 ml); ready for use. The vial contains less than 110 kBq of ^{125}I-labeled L-T$_4$ in buffer with protective proteins, sodium azide (< 0.1%) and yellow dye.

  2. Anti-T$_4$ Ab coated tubes; 100 tubes; ready for use.

  3. Standard: 6 vials (0.5 ml); ready for use. The vials contain solution of 0 to 20 µg/ dL of L-T$_4$ in human serum with sodium azide (< 0.1%).

  4. Control samples; 2 vial; lyophilized.

- **Method**

The method of L-T$_4$, which was determined, was based on the use of L-T$_4$ -RIA kit assay protocol as summarized in Table (2.3).

Table (2.3): Summary of the assay protocol of L-T$_4$.

Step 1 (Addition)	Step 2 (Incubation)	Step 3 (Washing and Counting)
• Adding sequentially to Ab-coated tubes • 20 µL of standard or sample or control and • 500 µL of tracer • Vortex	• Incubation for 1 hr (at 18-25°C) With • Shaking (at 280 rpm)	• Decant the contents of tubes carefully • Count activity (cpm) for at least one minute.

- **Calculation**

Mean counting, B/T ratio computation, and standard curve plotting were calculated similar to the calculation described in step 1, 2 and 3 of section 2.3, respectively. The L-T$_4$ concentration of the sample read off on horizontal axis in µg/dl (Figure 2.3).

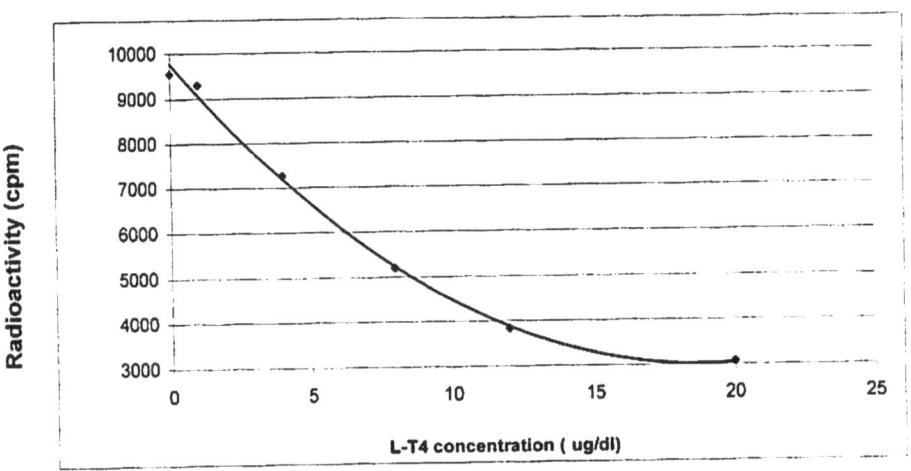

Figure (2.3): Standard curve of L-T$_4$ in human serum samples.

2.5. Determination of Total Triiodothyronine (L-T₃) Levels in Sera of Benign and Malignant Thyroid Tumor Patients and controls

Unknown samples or standard were incubated together with ^{125}I-L-T$_3$ in monoclonal anti-T$_3$ antibody coated tubes. After incubation the contents of the tubes were decanted and the bound activity was measured in a gamma counter. The concentration of L-T$_3$ was reversely proportionated to the radioactivity measured [141].

- **Reagents:**

  1. ^{125}I-triiodothyronine tracer; 1/5 per kit; Each vial containes 7 μci of tracer ($\leq$ 1 μg/ml T$_3$) in 6 ml of phosphate buffered saline with 0.1% sodium azide as preservative.

  2. Anti-T$_3$ Ab coated tubes; 100 tubes; ready for use.

  3. L-T$_3$ RIA assay buffer; 1/5 per kit; each bottle contains 120 ml of phosphate buffered saline 0.1% sodium azide as preservative.

  4. L-T$_3$ RIA serum standard: 5 vials (1.5 ml); ready to use. The vials contain solution of 0.5, 1.0, 2.5, 5.0 and 8.0 ng/mL respectively.

  5. L-T$_3$ RIA control sera levels I and II; 2 vial; lyophilized. The normal values for I = 1.5 $\pm$ 0.3 ng/ml and for II = 4.0 $\pm$ 0.8 ng/ml.

- **Method**

  The method of L-T$_3$ determination is based on the use of L-T$_3$ RIA kit assay protocol as summarized in Table (2.4).

Table (2.4): Summary of the assay protocol of L-T₃.

Step 1 (Addition)	Step 2 (Incubation)	Step 3 (Washing and Counting)
• Adding sequentially to Ab-coated tubes • 50 µL of standard or control or sample and • 1000 µL of tracer • Vortex	• Incubation for 1 hr at 37°C.	• Decant the contents of tubes carefully • Count activity (cpm) for one minute.

• **Calculation:**

Mean counting, B/T ratio computation, and standard curve plotting were calculated as described in steps 1, 2 and 3 of section 2.3, respectively. The L-T₃ concentration of the sample read off on horizontal axis in ng/ml (Figure 2.4).

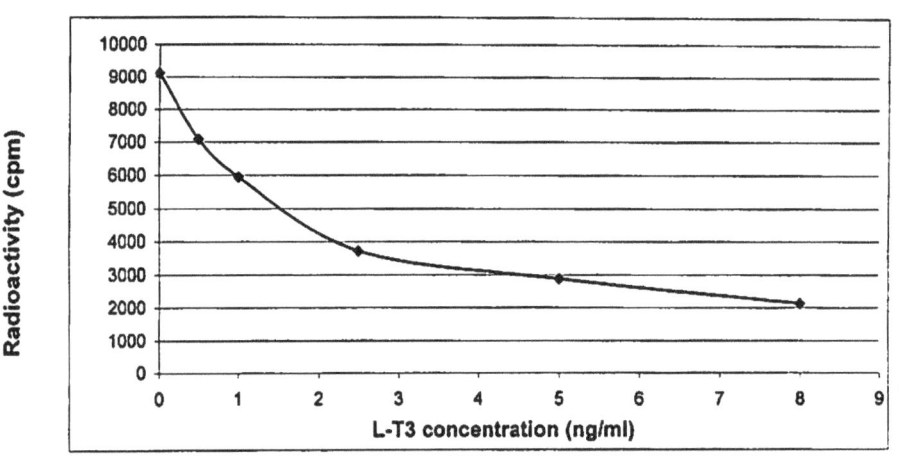

Figure (2.4): Standard curve of L-T₃ in human serum samples.

2.6. Determination of Total TSH Concentration in Benign and Malignant Thyroid Homogenates Tissues

The concentration of total TSH in benign and malignant thyroid tissue homogenates was determined by the same immunoradiometric assay (IRMA) used for serum total TSH determination [141] that was described in section (2.3).

- **Reagents:**
  1. The reagents provide for total TSH IRMA kit was described in section (2.3).
  2. The tissue homogenate was prepared as shown in section (2.1.7).
- **Method**

   The method for determination of total TSH in tissue homogenate was followed the assay protocol described in Table (2.2).

- **Calculation:**
  1. The TSH concentrations in IU/ml of benign and malignant thyroid tissue homogenates were estimated according to section (2.3).
  2. The IU/ml units were converted to mg/g tissue, since the tissue was homogenized in tris buffer with a ratio of 1: 3 (w/v) as mentioned in section (2.1.7).

2.7. Binding Studies of TSH in Benign and Malignant Thyroid Tissues with ^{125}I-Anti-Total Ab (Anti-tTSH Abs)

All the experiments of binding studies were carried out in duplicate.

2.7.1. Preliminary test of the binding of TSH in benign and malignant thyroid tissues with ^{125}Ianti-tTSH Abs

- **Method:**
  1. Twenty-five microliters (75 µg protein) of thyroid tissue homogenate was incubated with 25 µl (51 µg protein) of ^{125}I-anti tTSH Abs (mouse monoclonal IgG) at room temperature (25°C) for 120 minutes while continuously shaking.
  2. Two additional tubes containing 25µl ^{125}I-Ab only for total activity computation were set-aside until counting.

3. After incubation, the tubes were centrifuged at 500 xg and 4°C for 20 minutes in order to separate the TSH/anti-tTSH complex formed.

4. The supernatant was described be decanting the assay tubes. Then the tubes were inverted on a filter paper for 10 minutes.

5. The rims of the tubes were swabbed with a cotton wool and the amount of radioactivity (cpm) was counted in a gamma counter for 1 minutes.

- **Solution:**

  Tris buffer (pH = 7.8) was prepared as described in section (2.1.8).

- **Calculations:**

1. The counted radioactivity in each tube (expressed in cpm) represented the bound fraction (B) [i.e. TSH/anti-tTSH complex].

2. The counted radioactivity in the tubes containing labeled anti-tTSH Ab only represented the total activity (T).

3. The B/T ratio for each tube was counted as follows:

$$\frac{B}{T}\% = \frac{\text{Sample mean count (B)}}{\text{Total activity meancount (T)}} \times 100$$

2.7.2. Most appropriate conditions of the binding of TSH in benign and malignant thyroid tissues with ^{125}I-anti-tTSH Abs

2.7.2.1. The effect of different protein concentrations of benign and malignant thyroid tissue homogenates on binding of TSH with ^{125}I-anti-tTSH Ab

Methods, solutions and calculations were the same as described in section (2.7.1) except the following:

1. An increasing amounts (25, 50, 75, 100, 150, 200 and 250 μg) of thyroid tissue homogenates were incubated with 25 μl (51 μg) of ^{125}I-anti-tTSH Ab and the volume was made up to 250 μl with Tris buffer pH 7.4.

2. The percent of binding values B/T% were plotted against the increasing amounts of protein of the benign and malignant thyroid tissue homogenates (Figure 3.6).

2.7.2.2. The effect of different concentrations of ^{125}I-Anti-tTSH Ab on the binding with TSH in benign and malignant thyroid tissues

Methods, solutions and calculations were the same as described in section (2.7.1) except the following:

1. An increasing volumes (5, 7.5, 10, 12.5, 25 and 50 µl) which contain (10.2, 15.3, 20.4, 25.5, 51 and 102 µg protein) of ^{125}I-anti-tTSH Ab were incubated with 25 µl (75 µg) of thyroid tissue homogenate and the volume was made up to 250 µl with Tris buffer pH 7.4.

2. Another set of tubes containing the same increasing volumes of ^{125}I-anti-tTSH Ab only was used for total activity computation.

3. The time for centrifugation was 30 min instead of 20 min.

4. The percent of the binding values were plotted against the increasing volumes of ^{125}I-antitTSH Ab (Figure 3.5).

2.7.2.3. The effect of different pH values on the binding of TSH in thyroid tissue homogenates with ^{125}I Anti-tTSH Ab

Methods, solutions and calculations were the same as described in section (2.7.1) except the following:

1. Twenty-five microliters of thyroid tissue homogenate was incubated with 10 µl (20.4 µg) of ^{125}I-anti-tTSH Ab and the mixture was made up to 250 µl with Tris buffer of different pH values (6.8, 7.4, 7.8, 8.0, 8.2, 8.6, 9.0).

2. The percent of the binding values B/T% were plotted against the corresponding pH values (Figure 3.7).

2.7.2.4. Temperature dependency of the binding

Methods, solutions and calculations were the same as described in section (2.7.1) except the following:

1. Twenty-five microliters of thyroid tissue homogenate was incubated with 10 µl (20.4 µg) ^{125}I-anti-tTSH Ab and the mixture was made up to 250 µl with Tris buffer pH = 7.8.

2. The experiment was repeated at different temperatures (10, 25, 37 and 45°C) and the B/T% value was obtained at each temperature.

3. The percent of the binding values B/T% were plotted against different temperatures of incubation (Figure 3.8).

2.7.2.5. The choice of the most appropriate incubation time for the binding of TSH in thyroid tissue homogenates with ^{125}I Anti-tTSH Ab

Methods, solutions and calculations were the same as described in section (2.7.1) except the following:

1. Twenty-five microliters of thyroid tissue homogenate was incubated with 10 µl (20.4 µg) of ^{125}I-anti-tTSH Ab and the mixture was made up to 250 µl with Tris buffer pH = 7.8.

2. The assay mixtures were incubated at 25°C for benign and 45°C for malignant thyroid tissue homogenates at different time intervals (30, 60, 120, 150 and 180 minutes).

3. The percent of the binding values B/T% were plotted against different times of incubation (Figure 3.9).

2.7.2.6. Effect of different halides on the binding of TSH in thyroid tissue homogenates with ^{125}I Anti-tTSH Ab

- **Method**

1. Ten microliters (20.4 µg protein) of ^{125}I anti-tTSH Ab was incubated with 25 µl (75 µg protein) of thyroid tissue homogenate and the volume was made up to 250 µl with tris buffer pH 7.8 containing 0.1 M of each of the following halides (NaI, NaBr, NaF and NaCl).

2. The tubes were incubated for 180 min at 25°C for benign and at 45°C for malignant tumors.

3. At the same time a tube containing 10 µl (20.4 µg) of ^{125}I-anti-tTSH Ab only without the addition of any halide was used as a control.

4. After incubation the ^{125}I-anti-tTSH Ab/TSH complex was estimated as mentioned in section (2.7.1).

- **Solutions**

Halides solutions (0.1 M) were prepared by dissolving each of the following weights of halides in 5 ml of tris buffer pH 7.8; (1) NaI = 0.0799 g; (2) NaBr = 0.05144 g; (3) NaF = 0.02099 and; (4) NaCl = 0.029922 g.

- **Calculations**

1. The B/T% values were estimated as mentioned in section (2.7.1).

2. The binding percent was plotted against different halide (Figure 3.12-13).

2.7.2.7. Effect of divalent and monovalent cations on the binding of TSH in thyroid tissue homogenates with ^{125}I Anti-tTSH Ab

- **Method**

1. Ten microliters (20.4 µg protein) of ^{125}I anti-tTSH Ab was incubated with 25 µl (75 µg protein) of thyroid tissue homogenate and the volume was made

up to 250 µl with tris buffer pH 7.8 containing 25 mM of different divalent salts ($MgCl_2$, $MnCl_2$, $CuSO_4.5H_2O$, $CaCl_2$, $ZnCl_2$) and with monovalent 0.1 M of (LiCl, NH_4Cl, KCl).

2. The tubes were incubated for 180 min at 25°C for benign and at 45°C for malignant tumors.

3. At the same time a tube containing 10 µl (20.4 µg) of ^{125}I-anti-tTSH Ab only without the addition of any halide was used as a control.

4. After incubation the ^{125}I-anti-tTSH Ab/TSH complex was estimated as mentioned in section.(2.7.1).

- **Solutions**

The stock solutions of divalent salts (25 mM) and of monovalent salts (0.1 M) were prepared by dissolving each of the following weights of salts in 5 ml tris buffer pH 7.8;

(1) Divalent salts: [$ZnCl_2$= 0.017036 g; $CaCl_2$ = 0.013872 g; $CuSO_4.5H_2O$ = 0.03121 g; $MnCl_2.4H_2O$ = 0.02473 g; and $MgCl_2.6H_2O$ = 0.025413g].

(2) Monovalent salts: [LiCl = 0.021195 g; KCl = 0.037272 g; NH_4Cl = 0.026745 g].

- **Calculations**

1. The B/T% values were estimated as mentioned in section (2.7.1.3).

2. The binding percent was plotted against each salt (Figure 3.10-11).

2.7.3. The kinetics studies

All the experiments were carried out in duplicate.

2.7.3.1. The time course of the binding of TSH in benign and malignant thyroid tissue homogenates to ^{125}I-anti-tTSH Abs

Methods, solutions and calculations were the same as described in section (2.7.1) except the following:

1. Twenty-five microliters (75 µg protein) of thyroid tissue homogenate was incubated with 10 µl (20.4 µg) of ^{125}I-anti-tTSH Ab and the mixture was made up to 250 µl with Tris buffer pH = 7.8. The incubation was performed at different time intervals (30, 60, 90, 120, 150 and 180 minutes).

2. The experiment was performed at different temperatures (4, 10, 25, 37 and 45°C).

3. The percent of the binding values were plotted against different times of incubation at each temperature (Figure 3.15).

2.7.3.2. Determination of the affinity constant (K_a) and the maximal binding capacity (B_{max}) of TSH in benign and malignant thyroid tissue homogenates associated with ^{125}I anti-tTSH Abs

- Method

1. Twenty-five microliters (75 µg protein) of thyroid tissue homogenate were incubated with increasing volumes (2.5, 5, 7.5, 10 and 12.5 µl), which contain (5.1, 10.2, 15.3, 20.4 and 25.5 µg protein) of ^{125}I-anti-tTSH Ab. The incubation was carried out at 4°C. The final volume (250 µl) was made up by adding Tris buffer pH = 7.8.

2. Another set of tubes contain (2.5, 5, 7.5, 10 and 12.5 µl) of ^{125}I-anti-tTSH Ab only for the total activity computation was set-aside until counting.

3. After incubation the tubes were estimated the same as described in section (2.7.1).

4. The previous steps were also performed at different temperatures (10, 25, 37 and 45°C).

5. The time of incubation for benign and malignant tissue homogenates, needed to get the equilibrium state were as the following:

Temperature (°C)	Time (min)	
	Benign homogenate	Malignant homogenate
4	180	30
10	30	180
25	180	60
37	150	180
40	150	180

- **Solutions**

    The tris buffer (pH = 7.8) was prepared as described in section (2.1.8).

- **Calculation**

1. 1. The B/T ratio was computed for each tube where:

    B: Is the bound radioactivity counts (cpm), which represents the TSH/anti-TSH complex.

    T: Is the total activity mean counts.

    F: Is the free radioactivity mean counts (cpm) which represents the non-bound ^{125}I-anti-TSH Ab. It obtained by subtracting the B from T (i.e. F = T-B).

2. The concentration of ^{125}I-TSH/ anti-TSH complex in mg/ml that formed after time (t) was calculated from the following equation:

$$B(mg/ml) = \frac{B(cpm)}{T(cpm)} \times Conc. \text{ of } ^{125}I - Ab \text{ in the incubation medium in mg/ml}$$

3. The affinity constant and maximal binding capacity were determined according to Scatchard equation [142].

$$\frac{B}{F} = \frac{1}{k_d}(B_{max} - B)$$

$$k_a = \frac{1}{k_d}$$

Where:

K_a: The affinity constant.

K_d: The dissociation constant.

B_{max}: The maximal binding capacity.

4. The plot of B/F ratios vs. the B values in mg/ml gives a linear relationship. The value of K_a of binding at each temperature can be calculated from the slope of straight line, while the value of the total concentration of TSH (B_{max}) in benign and malignant thyroid tissues were calculated from the intercept with the x-axis.

2.7.3.3. The kinetic studies of TSH binding in benign and malignant thyroid tissues with ^{125}I-anti-tTSH Abs

Methods, solutions and calculations were the same as described in section (2.7.3.2), in addition to the following steps:

1. The foreword reaction rate constant (K_{+1}) which is also called complex formation rate constant was calculated from the following equation:

$$\ln\left[\frac{\left(^{125}I-ATSHAb/TSH\right)_e}{\left(^{125}I-ATSHAb/TSH\right)_e-\left(^{125}I-ATSHAb/TSH\right)_t}\right]$$

$$=k_{+1}t\left[\frac{\left(^{125}I-ATSHAb\right)_T(TSH)_T}{\left(^{125}I-ATSHAb/TSH\right)_e}\right]$$

Where:

- $\left(^{125}I\text{-A TSH Ab/TSH}\right)_e$: concentration of (^{125}I-anti TSH Ab/TSH) complex at equilibrium.

- $(TSH)_T$: total concentration of TSH.

- $\left(^{125}I\text{-A TSH Ab/TSH}\right)_t$: concentration of complex at time t.

- $\left(^{125}I\text{-A TSH Ab}\right)_T$: total concentration of ^{125}I-anti TSH Ab.

2. Backward reaction rate constant (k_{-1}) which is also called complex dissociation rate constant was calculated from the following equation:

$$ka = \frac{k_{+1}}{k_{-1}}$$

Where: k_a is the affinity constant.

2.7.3.4. The thermodynamic studies of TSH binding in benign and malignant thyroid tissues with ^{125}I-anti-tTSH Abs

• **Thermodynamic Parameters of Standard State**

  The thermodynamic parameters of standard state (ΔH°, ΔG° and ΔS°) were obtained from Van't Hoff plot. The value of the natural logarithm of equilibrium constant (affinity constant, k_a) obtained at different temperatures were plotted against the reciprocal values of absolute temperature in Kelvin (1/T) was calculated according to the following equation:

$$\ln k_a = \frac{\Delta S^\circ}{R} - \frac{\Delta H^\circ}{RT}$$

Where:

ΔH°: The enthalpy change of the standard state,

ΔS°: The entropy change of the standard state,

R: The gas constant (8.3144 J.mol^{-1}.K^{-1}).

ΔH° value was obtained from the slope of the linear relationship of the plot. The change in Gibbs free energy of the standard state (ΔG°) was obtained from the following equation:

$$\Delta G^\circ = -RT \ln k_a$$

The standard state entropy change (ΔS°) was calculated from the Gibbs equation:

$$\ln k_a = \frac{\Delta S^\circ}{R} - \frac{\Delta H^\circ}{RT}$$

- **Thermodynamic Parameters of Transition State**

    The thermodynamic parameters of the transition state were obtained from Arrhenius plot of $\ln k_{+1}$ versus $(1/T)$ that gives a linear relationship according to the following equation:

$$\ln k_{+1} = \ln A - \left(\frac{E_a}{RT}\right)$$

Where:

A: The Arrhenius constant,

E_a: The activation energy,

R: The gas constant, and

T: Absolute temperature in Kelvin.

The value of E_a of the binding reaction could be determined from the slope of the straight line. The enthalpy of the transition state $(\Delta H^{\bullet})$ was obtained from the following equation:

$$\Delta H^{\bullet} = E_a\text{-}RT$$

While the free energy of transition state $(\Delta G^{\bullet})$ was calculated by using the following equation;

$$\Delta G^{\bullet} = -RT \ln k_{+1} + RT \ln\left(\frac{kT}{h}\right)$$

Where:

k: is Boltzmann constant $(1.38 \times 10^{-23}$ J.deg$^{-1})$.

h: is plank constant $(0.662 \times 10^{-34}$ J.sec$^{-1})$.

The change in entropy of the transition state $(\Delta S^{\bullet})$ was calculated from the following equation:

$$\Delta S^{\bullet} = \frac{\left(\Delta H^{\bullet} - \Delta G^{\bullet}\right)}{T}$$

2.8. Isolation of ^{125}I-Anti TSH Ab/TSH complex and the Unbound of ^{125}I-Anti TSH Ab in Benign and Malignant Thyroid Tissue Homogenates

Gel filtration chromatography technique [(143)] was used for the isolation of ^{125}I-anti TSH Ab/TSH complex from unbound ^{125}I-anti TSH Ab in benign and malignant thyroid tissue homogenates.

2.8.1. Preparation of gel

The gel (sephadex G-150) was allowed to swell in excess of buffer (2 g of the gel in approximately 100 ml of the buffer) and left to stand for three days at room temperature without stirring to equilibrate with the buffer. The buffer was decanted and the gel was resuspended in excess volume of eluent of buffer three times before bed packing.

2.8.2. Bed packing

The de-gassed slurry was carefully mixed before pouring into the vertical column, which contains 5 ml of eluent buffer by using a glass rod, attached to the inner surface of the column. After the gel has been settled, the column outlet was opened packing was continued until the gel reached a stable bed height 30 cm the column was equilibrated with tris buffer for 24 hrs with dimensions 1×30 cm and a bed volume of 23.5 ml.

2.8.3. Void volume (Vo) determination

The void volume of the gel column was estimated by using blue dextran 2000 at concentration of (2 mg/ml) in deionized water 0.5 ml of blue dextran solution was carried out with the same buffer, using a flow rate of 10 ml/hr.

Fractions of 1 ml were collected and their absorbances were measured at 600 nm to determine the void volume.

2.8.4. Sample addition and protein radioactivity elution

- **Method**

One milliliter of the (^{125}I-anti TSH Ab/TSH) complex of benign and malignant thyroid tissues were prepared at their optimum conditions and applied to the column equilibrated with tris buffer. The fractions were eluted with same flow rate (10 ml/hr). The radioactivity was measured by Gamma counter and the absorbance for the eluted fractions were recorded at 600 nm.

- **Solutions**

1. A stock solution of tris buffer was prepared by dissolving (24.2 gm) of tris(hydroxy methyl amino methan) in (1000 ml) of distilled water, the required pH 7.8 for benign and malignant tumors was adjusted by adding HCl solution (0.2 M).

2. Sodium azide (0.02%) (w : v) had been added to certain components as antibacterial agent.

- **Calculations**

1. The radioactivity (cpm) and the absorbance were plotted vs. the number of fractions.

2. The fractions under each peak were pooled and the absorption spectrum was measured in the area (200-320 nm) by using a 0.5 cm cuvette against tris-buffer pH 7.8 for benign and malignant tissues in reference beam.

2.9. Spectroscopic Studies on TSH/^{125}I-Anti TSH Ab complex and the Unbound of ^{125}I-Anti TSH Ab

2.9.1. The UV-spectra of ^{125}I-anti TSH Ab/TSH complex and the unbound of ^{125}I-anti TSH Ab in benign and malignant thyroid tissues

• The uv-spectrum of ^{125}I-anti TSH/TSH complex and unbound of ^{125}I-anti TSH Ab

The gel filtration experiment in section (2.8) gave two peaks for each benign and malignant tissue homogenate groups. The first represents the ^{125}I-anti TSH Ab/TSH complex, while the second peak represents the unbound ^{125}I-anti TSH Ab (free fraction). The fraction under each peak was pooled.

• Method

♦ *Malignant thyroid tissue homogenate*

A volume of 100 μl (351 μg protein) of the pooling fraction of the complex peak and 100 μl (130 μg protein) of the pooling fraction of the free peek. Each of these was completed to 500 μl of tris buffer pH 7.8. Then placed in a 0.5 cm cuvette in a sample beam and the absorption spectrum was immediately measured against the same buffer in reference beam in the area of (200-350 nm).

♦ *Benign thyroid tissue homogenate*

The experiment was performed for benign tissue homogenate similar to that for malignant homogenates, but in case of benign the added volume (100 μl) represents 336μg protein and 30μg protein for polling fractions of the complex and free peaks, respectively.

- **Solution**

    Tris/HCl buffer was prepared as follows:

Solution A: (0.2 M) tris; 2.4228 g tris (hydroxy methyl amino methane) in 100 ml distilled water.

Solution B: (0.1 N) HCl.

Working Buffer (pH 7.8): Prepared by mixing 25 ml of solution A with an appropriate amount of solution B adjust the pH required, the volume was made up to 100 ml with distilled water.

2.9.2. Factors affecting the absorption properties of the ^{125}I-anti TSH Ab/TSH complex and the unbound of ^{125}I-anti TSH Ab in malignant thyroid tissue on UV-spectrum

2.9.2.1. pH effect
- **Method**

    The same was as described in section (2.9.1) but the volume was completed to 500 µl with different buffers at different pH values (4-12). In additions, the absorption spectrum was measured in the area (200-300 nm).

- **Solutions**

1. **Tris/HCl Buffer:** Prepared as described in section (2.9.1) but at different pH values (7.4, 7.8 and 9).

2. **Glycine/NaOH Buffer:** Prepared as follows:

    Solution A: (0.1 M) glycine in (0.1 N) NaCl; 0.7507 g glycine and 0.5844 NaCl dissolved in 100 ml distilled water.

    Solution B: (0.1 N) NaOH.

    Working Buffer pH (10.5, 12): Prepared by mixing appropriate amounts of solution A and B in a final volume of 100 ml.

3. **Acetate Buffer:** Prepared as follows:

Solution A: (0.1 M) sodium acetate; 0.8204 g $C_2H_3O_2Na$ in 100 ml distilled water.

Solution B: (0.1 N) acetic acid.

Working Buffer pH (4,6): Prepared by mixing appropriate amounts of solution A and B to reach the required pH in a final volume of 100 ml.

2.9.2.2. Polarity effect

A) The Effect of 20% of Methanol

One hundred microliter (351 µg protein) of the pooling peak (complex) in tube and 100 µl (130 µg protein) of the polling peak (free) in another tube were completed to 500 µl with tris buffer contains 20% methanol at pH 7.8, then each of which was placed in the test cell and the buffer containing 20% methanol was placed in the reference cell using 1 cm cuvette. The absorption spectrum of each sample was immediately measured.

B) The Effect of 20% of Chloroform

The same as described in (A) but instead the tris buffer contains 20% chloroform was used.

C) The Effect of 20% of Polyethylene glycol and 20% glycerol

The same as described in (A) but instead of the tris buffer 20% polyethylene or 20% glycerol were used.

D) The Effect of 20% Urea and 20% KCl

The same as described in (A) but instead the tris buffer contains 20% urea or 20% KCl were used.

2.9.3. The effect of NaCl concentration on thermal stability of the ^{125}I-anti TSH Ab/TSH complex and the unbound of ^{125}I-anti TSH Ab in malignant thyroid tissue on UV-spectrum

1. Two tubes of pooling fraction of malignant (complex and free) were prepared at two buffer solutions.

2. The tubes were heated for 15 min at 30°C.

3. The solution was placed in a 0.5 cm cuvette in a sample beam.

4. The absorption spectrum was immediately measured against the alkaline solution in reference beam in the area (220-350 nm).

5. The tubes were continuously heated for 15 min at six different temperatures (30, 40, 50, 60, 70 and 80)°C. The step 4 was repeated after each degree of heating.

- **Solutions**

  Buffer solutions were 20% ethylene glycol containing two different NaCl concentrations (0.1 M; 0.5844g, and 0.01 M; 0.0584 g) NaCl.

- **Calculations**

  The absorbance (A) was plotted vs. the temperature at two wavelength; λ_{max} = 292 nm; λ_{max} = 295 nm; using (0.1 M and 0.01 M) NaCl.

CHAPTER 3

Results & Discussion

Chapter Three

Results and Discussion

3.1. Determination of Thyroid Hormones and TSH in Sera of Thyroid Disease Patients and Normal Controls

hyroid hormones LT_3 and $L-T_4$, have long been recognized for their importance in regulating general metabolism, development and tissue differentiation [144-146]. In patients with thyroid nodule, the thyroid function test is mandatory to identify underlying thyroid pathology and not to differentiate benign from malignant nodules. The tests which are most widely used in clinical practice were serum immunoassay of LT_3, $L-T_4$ and TSH[147,148].

In the present work, the preoperative serum total LT_3, $L-T_4$ and TSH concentration were measured for patients with mutlinodular goiter (Group I), toxic goiter (Group II), and thyroid carcinoma (Group III). The three groups with a group of control subjects (Group IV).

Table (3.1) shows the mean ± SD and the ranges of LT_3, $L-T_4$ and TSH concentration as well as Figure (3.1-3.3) illustrates their distribution study reveals a significant increase ($P < 0.05$) in the mean levels of serum LT_3 (4.46 ± 0.75 ng/ml) and serum T_4 (16.86 ± 1.55 µg/dl), that associated with a significant decrease in the mean serum TSH levels (0.09 ± 0.046 IU/ml) of patients with toxic goiter in comparison to normal controls (1.1 ± 0.17 ng/ml, 7.92 ± 1.12 µg/dl) and (1.65 ± 0.4 IU/ml respectively). While no significant differences were observed, for the three studied hormones, between normal controls and each of multinodular goiter group (1.17 ± 0.24 ng/ml, 8.0 ± 1.4µg/dl and 1.68 ± 1.17

56

IU/ml respectively) and thyroid cancer patients (1.3 ± 0.4ng/ml, 7.75 ± 3.2 μg/dl and 1.76 ± 1.075 IU/ml) respectively.

Table (3.1): Thyroid Hormones and TSH concentration in sera of thyroid Disease patients and normal controls (mean ± SD).

Serum thyroid tests	Studied groups $\overline{X}$ ± SD (range)			
	Multinodular goiter	Toxic goiter	Thyroid carcinoma	Normal controls
L-T₃ (ng/ml)	1.17 ± 0.24* (0.7-1.5)	4.46 ± 0.75 ** (3.4-6.0)	1.3 ± 0.4* (0.9-2.2)	1.1 ± 0.17 (0.9-1.4)
L-T₄ (μg/dl)	8.0 ± 1.4* (4.7-10.4)	16.86 ± 1.55** (14.3-19.8)	7.76 ± 3.2* (4-12.4)	7.92 ± 1.12 (6.8-9.6)
TSH (IU/ml)	1.68 ± 1.17* (0.1-4.7)	0.09 ± 0.046** (0.02-0.2)	1.76 ± 1.075* (0.1-3.6)	1.65 ± 0.4 (0.9-2.5)
* P value > 0.05 in comparison to normal controls.				
** P value < 0.05 in comparison to normal controls.				

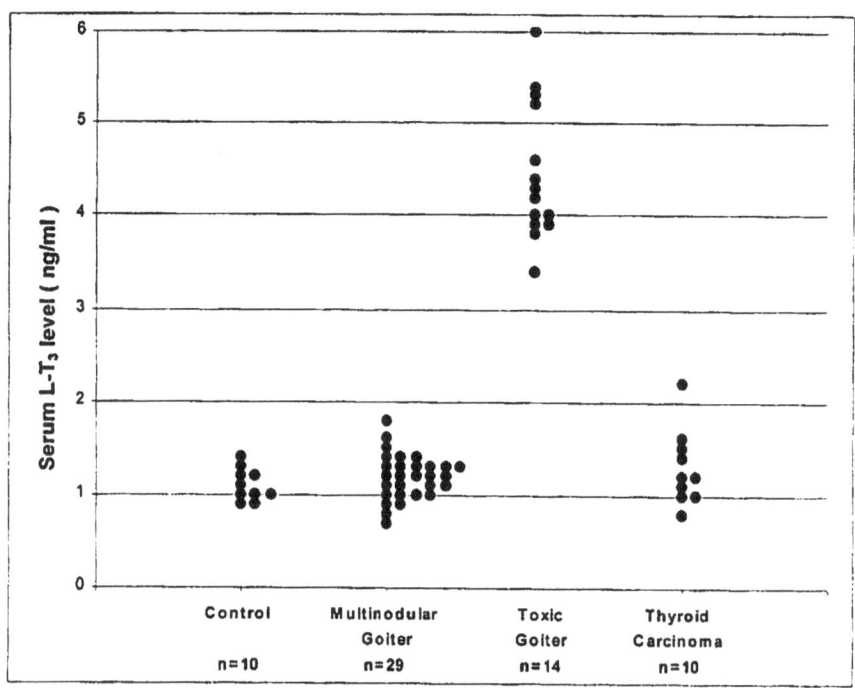

Figure (3.1): Distribution of serum L-T₃ levels of thyroid disease groups in comparison to normal controls.

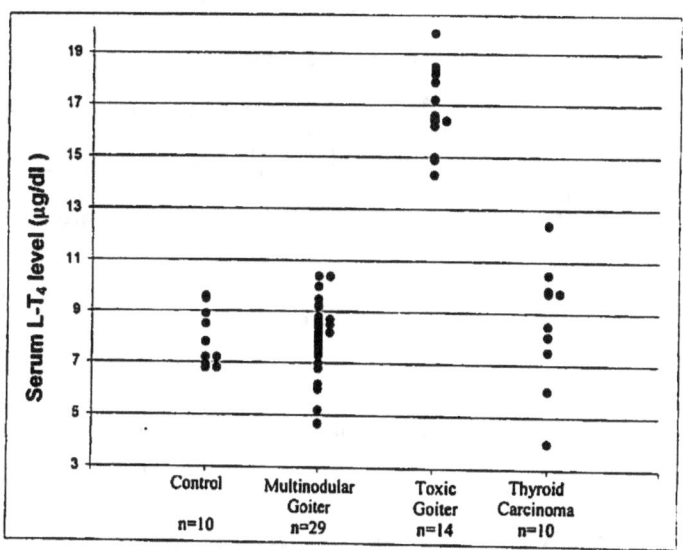

Figure (3.2): Distribution of serum L-T₄ levels of thyroid disease groups in comparison to normal controls.

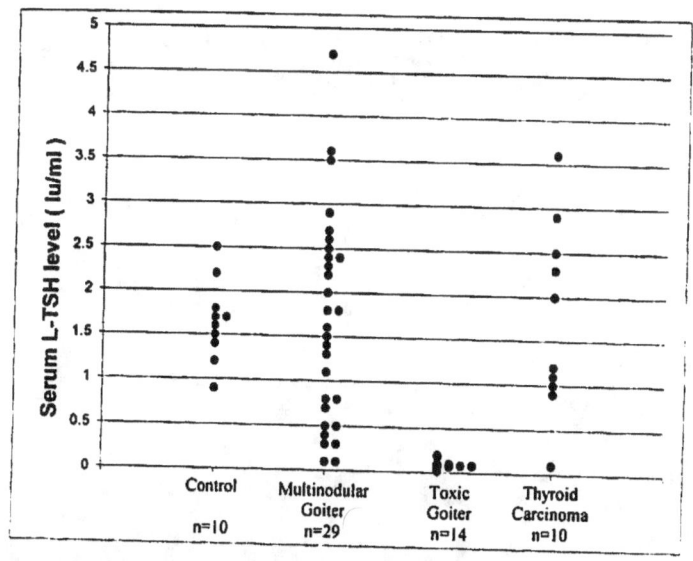

Figure (3.3): Distribution of serum TSH levels of thyroid disease groups in comparison to normal controls.

3.2. Thyroid Histopathology

The majority of thyroid diseases including carcinoma presented as asymptomatic thyroid nodule [26]. Whatever the presentation, histopathological examination of thyroid biopsy was the best to distinguish between benign and malignant nodules as well as it was of supreme importance for therapeutic purposes [149]. In addition to that, histopathological picture gives a hint to the functional status of the gland, but in such cases, thyroid function test was mandetored [150].

Figure (3.4a-F) demonstrated the histopathological pictures of different thyroid diseases chosen in the present work. Interestingly, a significant difference could be shown in different diseases, this could be further support biochemical studies on tissue in addition to that in serum.

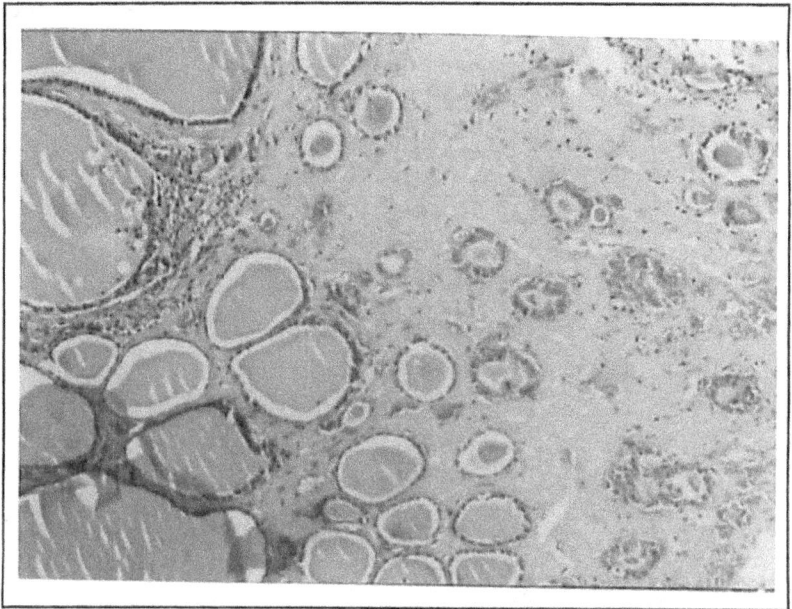

Figure (3.4a): Multinodular goiter; histopathological examination; non-toxic areas, active areas with fibrosis separating the nodules. The size of the follicules was of varying diameter.

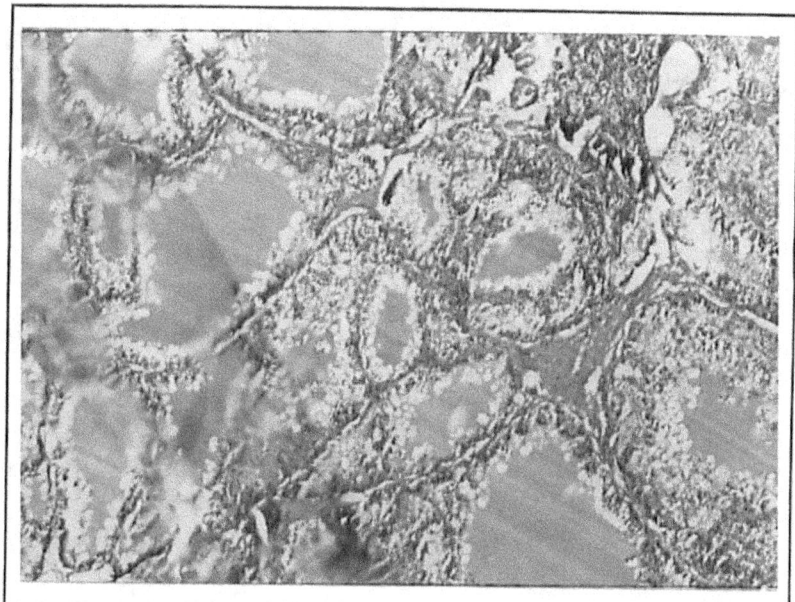

Figure (3.4b): Grave's disease; histopathological examination; diffuse thyroid hyperplasia, with hyper plastic, follicles peripheral vaculation of colloid and columinar cell lining. Scattered lymphocytic accumulation were noted.

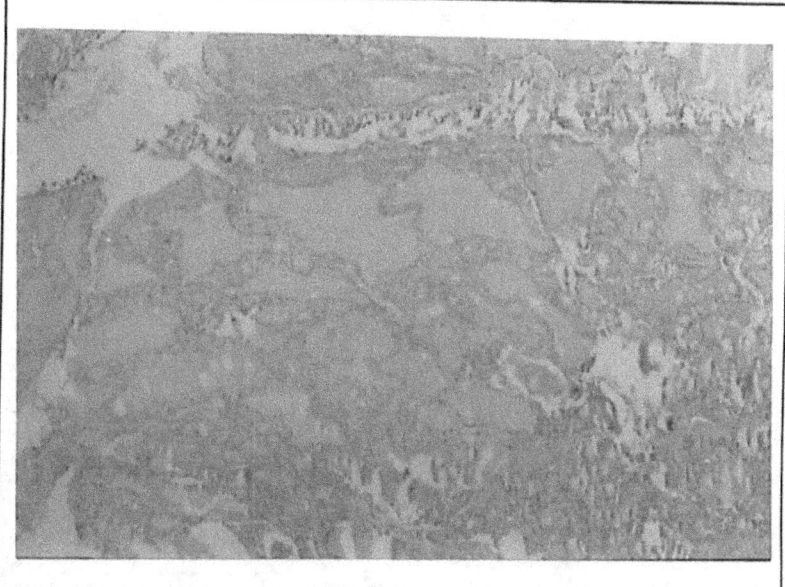

Figure (3.4c): Toxic nodular goiter; histopathological examination; sometimes toxic nodule is present in nodular goiter can lead to hyperthyroidism.

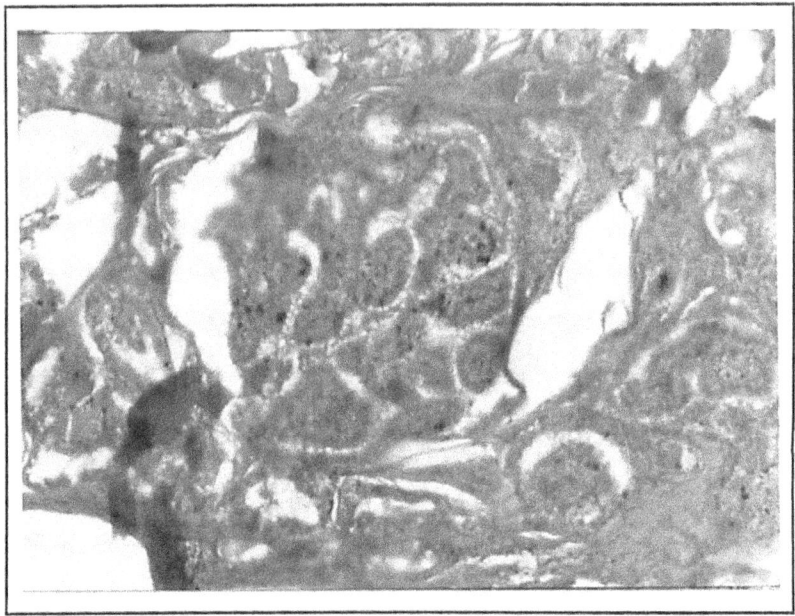

Figure (3.4d): Papillary thyroid carcinoma; histopathological examination; papillary projections with crowding of nuclei which are large and pale with psammoma bodies in 50% of the cases.

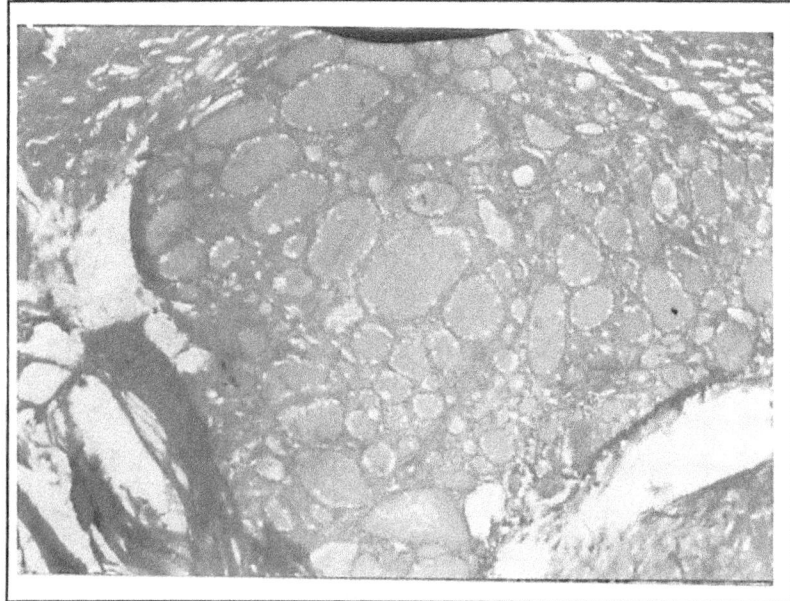

Figure (3.4e): Follicular thyroid carcinoma; histopathological examination; abnormal follicular structures with capsular invasion. Well-differentiated follicular carcinoma of the thyroid.

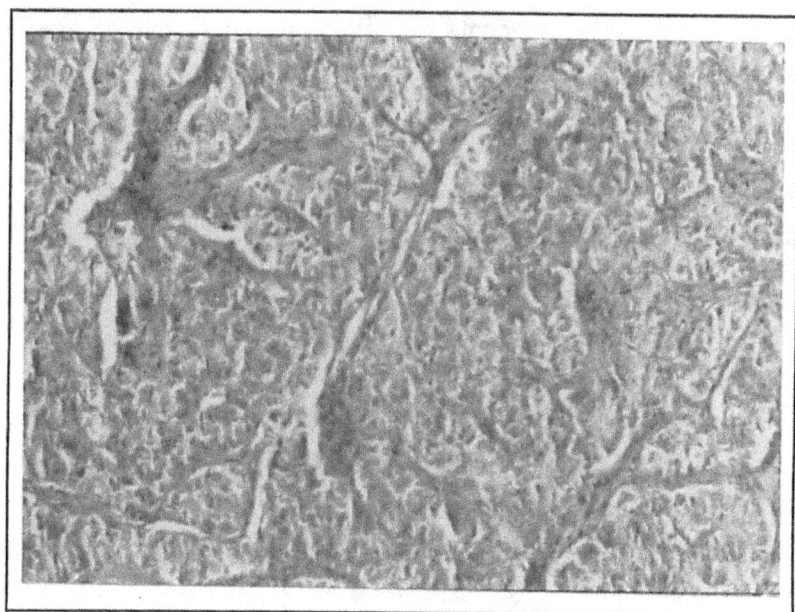

Figure (3.4f): Anaplastic thyroid carcinoma; histopathological examination; masses of undifferentiated separated by fibrosis.

3.3. Binding Studies of TSH in Benign and Malignant Thyroid Tissues with ^{125}I-Anti TSH Ab

Human thyroid taken throughout this study were inflicted with benign and malignant thyroid tissues. As confirmed by histopathological examination, two groups of patients were included in this study. Group I consisted of 43 patients with benign thyroid tissue homogenates, their mean age was (33.1 ± 12.2 years) with age range (10-58 years). While group II consisted of 10 patients with malignant thyroid tissue homogenates, their mean age was (36.4 ± 14.1years) with age range (18-64 years).

The weights of resect tissue samples ranged between (1-9) grams. The homogenization and sonication of the samples were carried out in a cold medium (4°C) to avoid protein denaturation and to decrease the proteolytic enzyme activity.

The tissue homogenate was filtered through several layers of nylon gauze to remove any suspended pieces of unhomogenized tissue or blood vessels, while centrifugation at 4000 xg removed the unruptured cells and intact nuclei of the ruptured cells leaving other cytoplasmic constituents in the supernatant, which was used as a source of TSH in the present study.

The amount of protein was determined by Lowry's method [139].

3.4. Determination of TSH Concentration in Benign and Malignant Thyroid Tissues by IRMA

Tissue concentration of TSH was determined by immunoradiometric assay (IRMA) in:

- Single sample of benign thyroid tissue homogenate, the mean of TSH concentration was 0.5mg per gram fresh tissue.

- A pool of benign thyroid tissue homogenates of 19 patients (Group I), the mean concentration was 0.452 mg per gram tissue, and a pooled of malignant thyroid tissue homogenate of the 10 patients (Group II), the mean concentration was 0.125 mg per gram.

3.5. Binding Studies of TSH in Benign and Malignant Thyroid Tissue Homogenates with ^{125}I-anti tTSH Ab

3.5.1. Preliminary test of the binding of TSH in benign and malignant thyroid tissue homogenates with ^{125}I-anti tTSH Ab

Benign and malignant thyroid tissue homogenates were used as the source of TSH in this study. The homogenate in benign or malignant was incubated with ^{125}I-anti tTSH Ab (mouse monoclonal IgG) for 120 min. The ^{125}I-AbAg complex formed was separated from the unbound particulates by centrifugation at 1500 xg for twenty minutes, this centrifuge speed was sufficient to precipitate the complex. After centrifugation, the tubes were decanted in order to get rid of the unbound Ab or Ag present in supernatant fraction while the ^{125}I-AbAg complexes remained as a pellet in the bottom of the tube.

The preliminary condition used in this experiment resulted in 15.3% binding in benign and 13.2% binding in case of malignant. The higher protein concentration in benign tissue homogenate seems to be the cause of relatively higher percent of binding tissue homogenate than that of malignant tissue homogenate [151].

3.5.2. Most appropriate condition of the binding of TSH in benign and malignant thyroid tissue homogenates with ^{125}I-anti TSH Ab

3.5.2.1. The effect of different concentrations of ^{125}I-Anti TSH Ab on the binding with TSH in benign and malignant thyroid tissue homogenates

One of the factors that affect the binding of Ab-Ag reaction is the concentration of the Ab fixed amounts of benign or malignant thyroid tissue homogenates [94] were incubated with increasing amounts of ^{125}I-anti tTSH Ab (10.2, 15.3, 20.4, 25.5, 51 and 102 µg). The protein determination was carried out according to Lowry's method [139] (section 2.2). The incubation was carried out at room temperature for 120 min.

The results revealed that the percent of binding in benign or malignant tissue homogenates increased by the volume (5 and 7.5 µl) until reached 10 µl and then decreased in the (12.5, 15, 25 and 50 µl) of ^{125}I-anti tTSH Ab added.

Figure (3.5) represented the ^{125}I-anti tTSH Ab curve with TSH in benign and malignant thyroid tissue homogenates. As shown in the Figure, the ^{125}I-Ab saturated with TSH molecules at the concentration of 20.4 µg/ml in tissue homogenates of both benign and malignant (Figure 3.5). Accordingly, in all subsequent, 10 µl (20.4 µg/ml) was used, since it gives the highest binding.

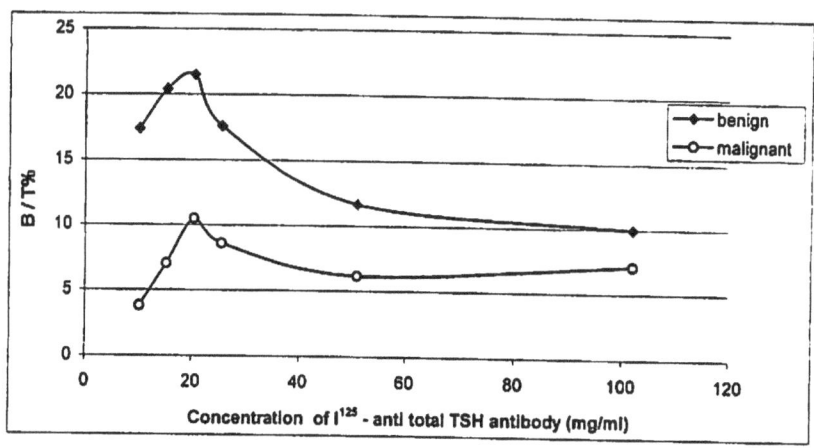

Figure (3.5): The effect of different concentration of ^{125}I-anti total TSH antibody on the binding with TSH in Benign and Malignant thyroid tissue homogenates. Details are described in section (2.7.2.2).

3.5.2.2. The effect of different protein concentrations of benign and malignant thyroid tissue homogenates on the binding of TSH with ^{125}I-Anti TSH Ab

To determine whether the different protein concentrations of benign or malignant affects the binding, an increasing amounts of homogenates (25, 50, 75, 100, 150, 200 and 250 µg) were incubated with 10 µl (20.4 µg/ml) of ^{125}I-anti tTSH Ab (according to the details in section (2.7.2.1).

Figure (3.6) represents the quantitative precipitation curve in which the amounts of (Ab-Ag) complex that precipitate were plotted as a function of Ag concentration. As shown in this Figure, in the first phase of reaction no precipitate was formed as increasing amounts of Ag were added. The amount of precipitate increased until a point of maximum binding was reached. After this point, as the amount of Ag increased, the amount of precipitate diminished. The increase in protein concentration which would increase the number of binding site and hence increase the percent of binding until reach the saturation state at 75 µg/ml homogenate concentration.

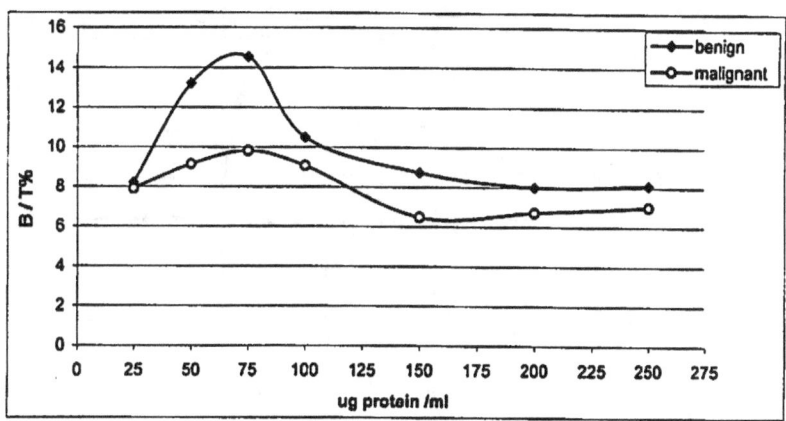

Figure (3.6): Effect of increasing amounts of Benign and Malignant thyroid tissue homogenates with ^{125}I-anti total TSH antibody. Details are described in section (2.7.2.1).

The (Ab-Ag) complex precipitate out of solution because of the multivalent nature of both molecules [94]. The radioactive Ab (IgG) was directed against a single TSH molecule and since the IgG Ab has two binding sites, it can cross-link antigenic sites of two different TSH molecules and form a lattice of interlocking molecules. As the size and complexity of the lattice increase, it becomes insoluble and precipitation out of solution. In all subsequent experiments, an amount of (75 µg/ml) protein of tissue homogenate was used according to the results obtained in this experiment.

3.5.2.3. The effect of different pH values on the binding of TSH in benign and malignant thyroid tissue homogenates with ^{125}I-anti tTSH Ab

The analysis of the influence of pH on binding of TSH in benign and malignant thyroid tissue homogenates to ^{125}I-anti tTSH Ab were stated in Figure (3.7).

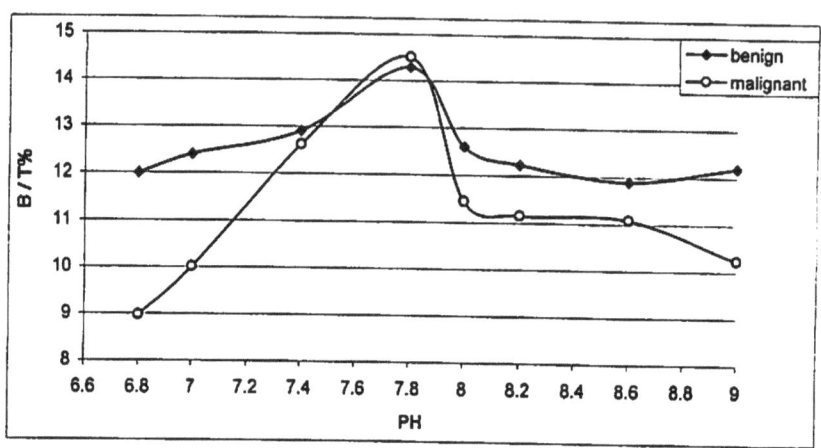

Figure (3.7): The effect of different pH on the binding of ^{125}I-anti TSH antibody with TSH in   Benign and Malignant thyroid tissue homogenates. Details are described in section (2.7.2.3).

The optimum pH was found to be 7.8 for the binding in benign and malignant homogenates, these results indicates that the binding was pH dependent. This shift in the pH of the environment may includes the induction of the protonation-deprotonation process [152] occurring within the charged polar groups on the amino acid residues present in the binding domain. According to the results obtained in this experiment, the pH of incubation buffer in all subsequent steps adjusted to pH 7.8 in benign or malignant since it gives maximum binding.

3.5.2.4. Temperature dependency of the binding

The temperature dependency of the binding of ^{125}I-anti tTSH Ab to benign and malignant thyroid tissue homogenates were investigated. Figure (3.8) represents the results of this analysis in benign tissue homogenate, it seems that the binding was the same when the temperature raised from 4°C to 10°C and the binding was reached a maximum at 25°C and then decreased at 37°C and a little raised at 45°C. The decrease in the binding may be due to the increase of (Ab-Ag) complex solubility [94], as a result the amount of bound fraction was diminished at high temperatures. According to these results 25°C was used in all subsequent experiments of (Ab-Ag) binding studies for benign thyroid tissue homogenate. On the other hand, figure (3-8) shows the result of malignant thyroid tissue homogenate, it seems that the binding was increased when the temperature was raised from 4°C to 10°C and then the binding decreased at 25°C and 37°C and then raised and reached maximum binding at 45°C. the increase in the binding may be due to the decrease of (Ab-Ag) complex solubility [94], as a result the amount of bound fraction was increased at high temperature. According to the results 45°C was used in all subsequent experiments of (Ab-Ag) binding studies.

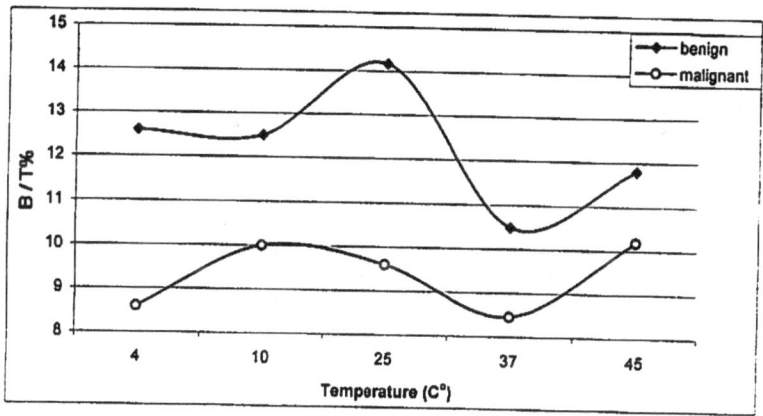

Figure (3.8): The effect of temperature on the binding of ^{125}I-anti TSH in benign and malignant thyroid tumors homogenate. Details are described in section (2.7.2.4).

3.5.2.5. The Choice of the Most Appropriate Incubation Time for the Binding of ^{125}I-Anti tTSH Ab with TSH in Benign and Malignant Thyroid Tissue Homogenates

To choose the most appropriate incubation time at 25°C for benign and 45°C for malignant thyroid tissue homogenates, the experiment was carried out at different time intervals (30-180 min). Figure (3-9A) shows that the optimal binding of ^{125}I-Ab to TSH in benign tissue homogenate was occurred within 3 hours and that the binding rise until 180 min when reached the optimum. In view of these results, the incubation time used in all subsequent experiments was 3 hours. On the other hand, Figure (3-9B) shows that the optimal binding of ^{125}I-Ab to TSH in malignant thyroid tissue homogenate was occurred also within 3 hours, while binding was decreased after this time. In view of these results, the incubation time used in all subsequent experiments was 3 hours. Similar results have been observed by other investigators [94] who found that the total binding of ^{125}I-TSH Ab to TSH in thyroid tissue homogenate was dependent on incubation time.

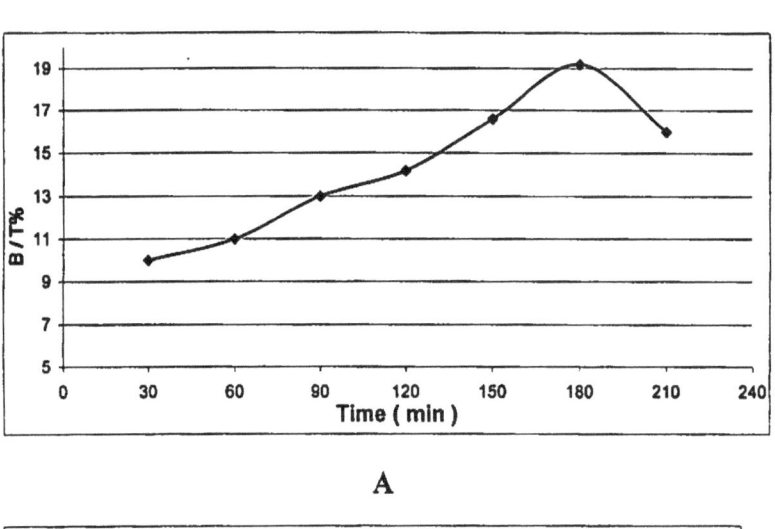

A

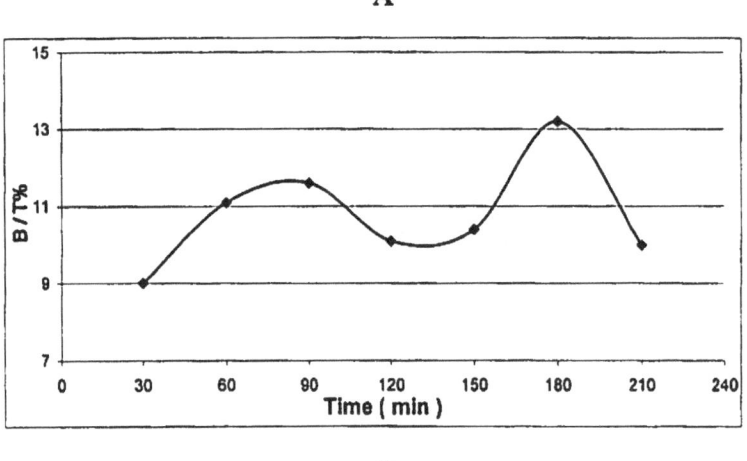

B

Figure (3.9): Time course of ^{125}I-anti total TSH antibody binding with TSH in, A) Benign and, B) Malignant thyroid tissue homogenate. Details are described in section (2.7.2.5).

3.5.2.6. The effect of divalent and monovalent cations on the binding of TSH in benign and malignant thyroid tissue homogenates with ^{125}I-Anti TSH Ab

Figure (3.10) shows the effect of mono and divalent on the binding of benign thyroid tissue homogenate to ^{125}I-anti tTSH Ab. The presence of NH_4^+ and K^+ at 0.1 M concentration seems to inhibit the binding and also all the other divalent cations did the same thing except $CaCl_2$. The interaction of these ions with the ionic groups of the (AbAg) complex diminishes the (AbAg) interaction and therefore, increasing solubility of the complex (salting in phenomenon). While the 0.1 M of LiCl and 25 mM of $CaCl_2$ were able to increase the binding to 9.7% in LiCl and 9.2% in $CaCl_2$. The solvent molecules were bound so tightly by the cations, that was unlike to solvate the Ab-Ag complex; the solute came out of solutions (salting out phenomenon).

In addition to that, Figure (3-11) shows the effect of mono- and divalent cations on binding of malignant thyroid tissue homogenate to ^{125}I-anti tTSH Abs. All the mono and divalent cations inhibit the binding. This is due to the increasing of (Ab-Ag) complex solubility in the presence of these cations. The interaction of these ions with the ionic groups of (AbAg) complex diminishes the (AbAg) interactions and therefore increasing solubility of the complex.

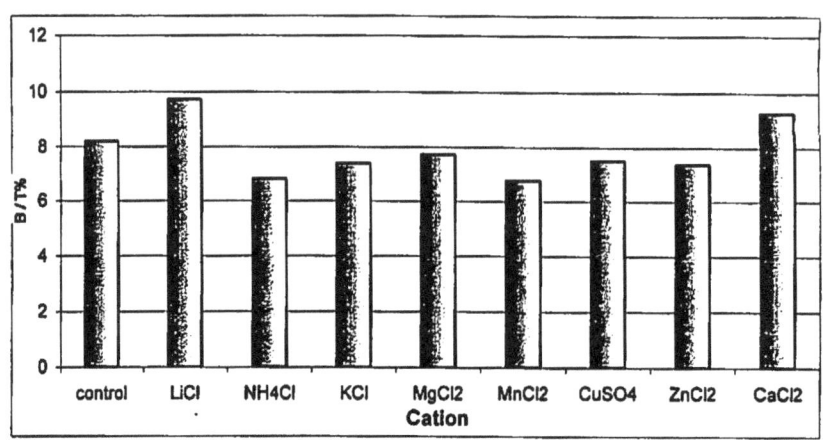

Figure (3.10): The effect of mono and divalent cations on the extent of binding of ^{125}I-anti total TSH antibody with TSH in Benign thyroid Homogenate tissue at 25mM and 0.1 M. Details are described in section (2.7.2.7).

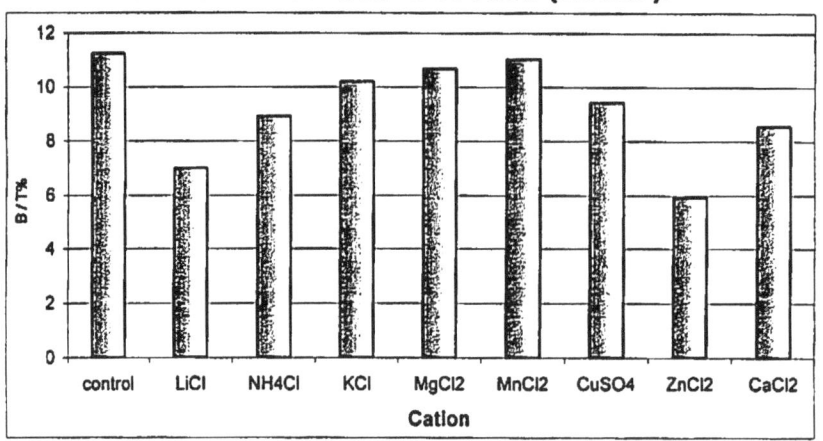

Figure (3.11): The effect of mono and divalent cations on the extent of binding of ^{125}I-anti total TSH antibody with TSH in Malignant thyroid Homogenate tissue at 25 mM and 0.1 M. Details are described in section (2.7.2.7).

3.5.2.7. The effect of different halides on the binding of TSH in benign and malignant thyroid tissue homogenates with ^{125}I-anti TSH Ab

Figure (3-12) shows the effect of different halide salts (i.e, NaI, NaBr, NaF, NaCl) at 0.1 M concentration on the extent of binding ^{125}I-anti tTSH Ab to benign thyroid tissue homogenate. It seems that the sodium halides inhibited the Ab-Ag binding according to the following order: NaF > NaCl = NaBr > NaI. The order corresponds to the increasing ionic radius and decreasing radius of hydration. Presumably, the lesser degree of hydration permits greater interaction of the salt with an ionic group located in the Ab or Ag combining sites.

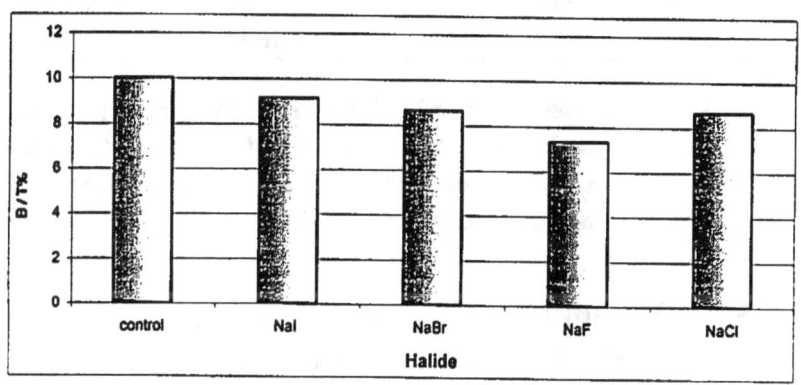

Figure (3.12): The effect of different halides on the extent of binding of ^{125}I-anti total TSH antibody with TSH in benign thyroid homogenates tissue at 0.1M. Details are described in section (2.7.2.6).

On the other hand, Figure (3-13) shows the effect of different halide salts (i.e. NaI, NaF, NaBr and NaCl) at 0.1 M concentration on the extent of binding of ^{125}I-anti tTSH Ab to malignant thyroid tissue homogenate, it seems that NaI, NaBr and NaF inhibited the Ab-Ag binding according to NaI > NaBr > NaF while only NaCl increasing the Ab-Ag binding. The cause that NaI, NaBr and NaF inhibit complex binding in this order may correspond to the decreasing ionic radius and increasing radius of hydration. The leaser degree of hydration

72

permits greater interaction of the salt with an ionic group located in the Ab or Ag binding sites.

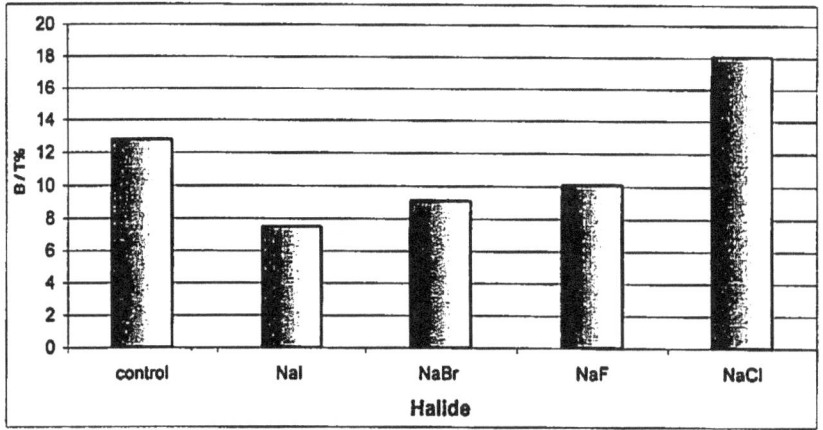

Figure (3.13): The effect of different halides on the extent of binding of ^{125}I-anti total TSH antibody with TSH in malignant thyroid homogenate tissue at 0.1M. Details are described in section (2.7.2.6).

3.5.3. Determination of the affinity constant (K_a) and the maximal binding capacity (B_{max}) of TSH in benign and malignant thyroid tissue homogenates the associated with ^{125}I-anti tTSH Ab

The concentration of TSH in benign and malignant thyroid tissue homogenates (B_{max}) and the affinity constant (Ka) of the binding to anti tTSH Ab has been measured. The experiment was carried out at the optimal conditions that were obtained in the previous experiments.

Scatchard plot analysis [142] in both benign and malignant homogenates gave straight lines as shown in Figure (3.14).

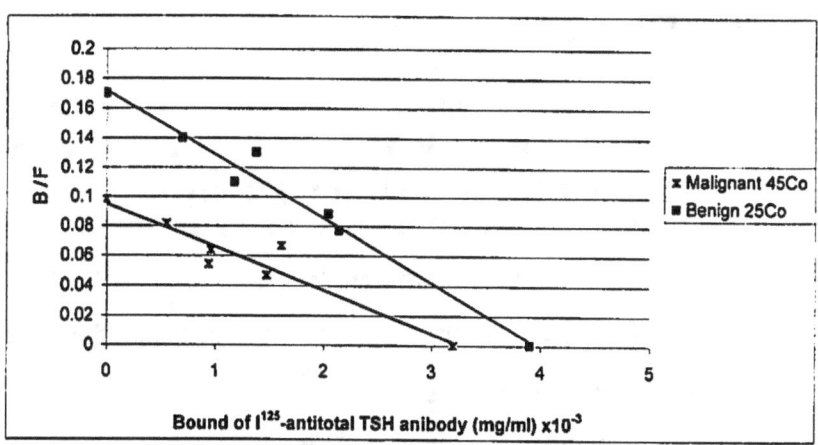

Figure (3.14): Scatchrad plot of ^{125}I-anti TSH antibody binding with TSH in Benign and Malignant thyroid homogenate tissues. Details are described in section (2.7.3.2).

The results obtained indicate that ^{125}I-Ab directed against the same epitope on the f-TSH molecule.

The affinity constants of the binding were (60, 99 mg^{-1}.ml) and the tissue concentration of TSH were (3.9, 3.2mg/ml) of benign and malignant thyroid tissue homogenates respectively at their optimum conditions.

Since the values of TSH concentrations in benign and malignant thyroid tissue homogenates obtained from this experiment and the one obtained by IRMA (section 2.6) were so close to each other. Hence, it is possible to use Scatchard analysis to measure TSH level in thyroid tissues.

3.6. The Kinetic Studies

3.6.1. The kinetics of the interaction of ¹²⁵I-anti tTSH Ab with TSH in benign and malignant thyroid tissue homogenates

Figure (3.15) shows the time course of the formation of (^{125}I-anti tTSH/TSH) complex at five different temperatures (4, 10, 25, 37 and 45°C). The concentration of the (^{125}I-anti tTSH/TSH) complex that formed after time (t) was calculated from the following equation:

$$B(mg/ml) = \frac{B(cpm)}{T(cpm)} \times \text{concentration of}\ ^{125}I - Ab\ \text{in the incubation medium in mg/dl}$$

The results of the time course pattern at different temperatures revealed that the binding of ^{125}I-anti tTSH Ab to TSH in benign and malignant thyroid tissue homogenates was a temperature dependent process with a maximum binding occurs at 25°C and 3 hours in benign and at 45°C and 3 hours in malignant.

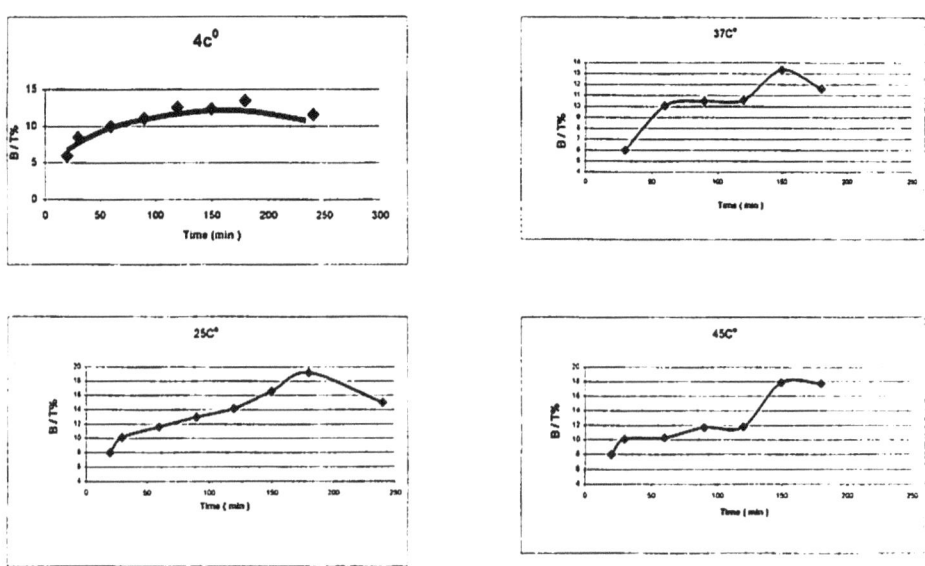

Figure (3.15a): Time course of the binding of TSH with ¹²⁵I-anti TSH antibody in benign thyroid homogenate tissue. Details are described in section (2.7.3.1).

75

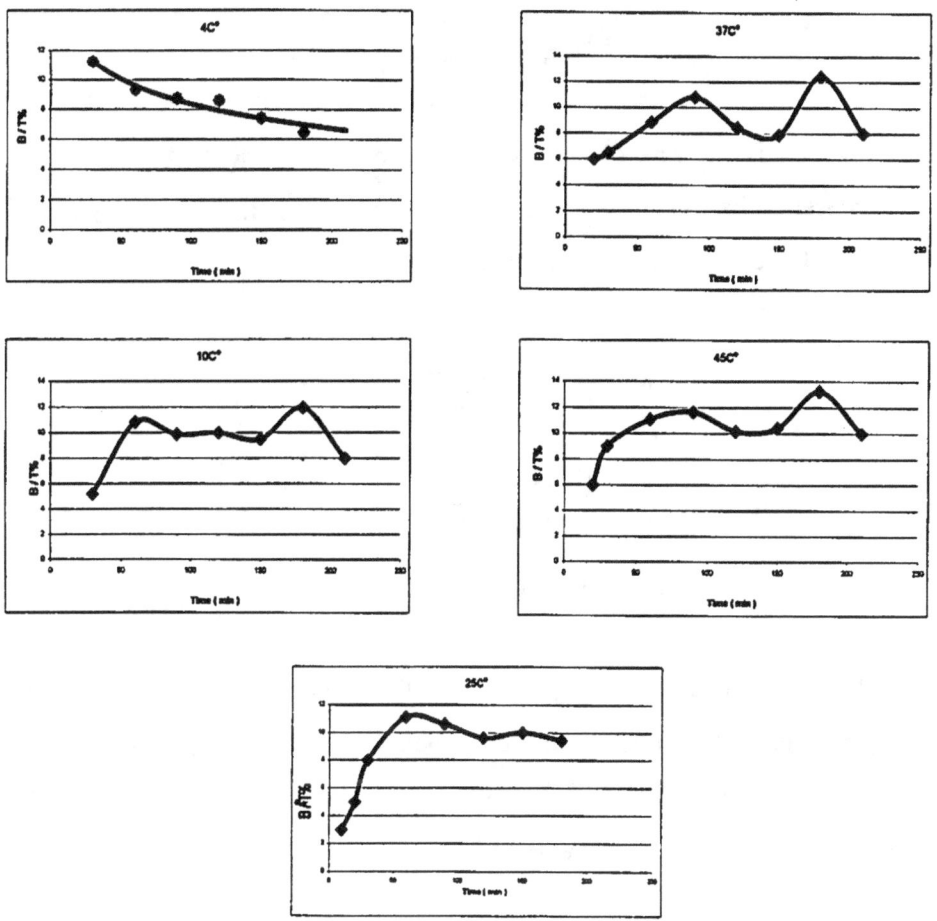

Figure (3.15b): Time course of the binding of TSH with ^{125}I-anti TSH antibody in malignant thyroid homogenate tissue. Details are described in section (2.7.3.1).

3.6.2. Determination of the kinetics parameters of the ^{125}I-anti tTSH Ab binding with TSH in benign and malignant thyroid tissue homogenates

The time course of ^{125}I-anti tTSH Ab binding to TSH in benign and malignant thyroid tissue homogenates was carried out to describe the kinetic parameters of the binding. The simplest proposed model representing the interaction of ^{125}I-Ab with TSH could be expressed by the following equation:

$$^{125}\text{I-Ab} + \underset{(\text{TSH})}{\text{Ag}} \underset{k_{-1}}{\overset{k_{+1}}{\rightleftharpoons}} {}^{125}\text{I-AbAg}$$

k_{+1} is the rate of the association of ^{125}I-Ab with Ag (or TSH) and k_{-1} represents the rate of the reverse reaction of the dissociation of the complex formed under the same conditions:

At equilibrium:

$$k_a = \frac{[^{125}\text{I} - \text{AbAg}]}{[^{125}\text{I} - \text{Ab}][\text{Ag}]} \quad \dots\dots\dots\dots\dots \quad (1)$$

$$k_d = \frac{[^{125}\text{I} - \text{Ab}][\text{Ag}]}{[^{125}\text{I} - \text{AbAg}]} \quad \dots\dots\dots\dots\dots \quad (2)$$

$$k_a = \frac{1}{k_d} = \frac{k_{+1}}{k_{-1}} \quad \dots\dots\dots\dots\dots\dots \quad (3)$$

Where k_a is the equilibrium constant of the association (affinity constant) and k_d is the equilibrium constant of the dissociation of ^{125}I-AbAg complex.

The values of k_a and B_{max} were calculated from Scatchard plot [142] at five different temperatures in Table (3.2) and Figure (3.16).

Table (3.2): The kinetic parameters of ^{125}I-anti tTSH Ab binding to TSH in benign and malignant thyroid tissue homogenates. All other details are described in section (2.7.3.2).

Temp. (°C)	Binding capacity × 10^-3 (mg/ml protein)		$k_d = k_{-1}/k_{+1} \times 10^{-2}$ mg/ml		$k_a = k_{+1}/k_{-1} \times$ mg^-1.ml	
	Benign	Malignant	Benign	Malignant	Benign	Malignant
4	2.49	2.61	2.1	1.9	46	51
10	2.47	2.74	1.8	1.0	53	93
25	3.9	3.14	1.6	1.4	60	68
37	2.5	2.88	1.7	1.6	57	59
45	2.8	3.2	1.8	1.0	55	99

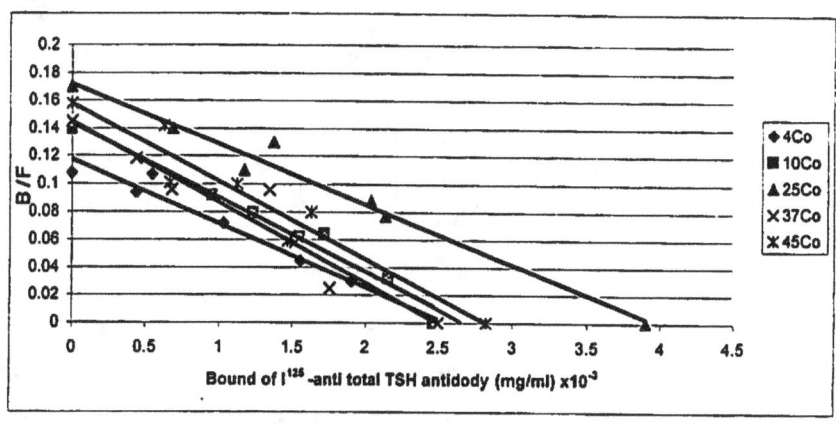

A: Benign

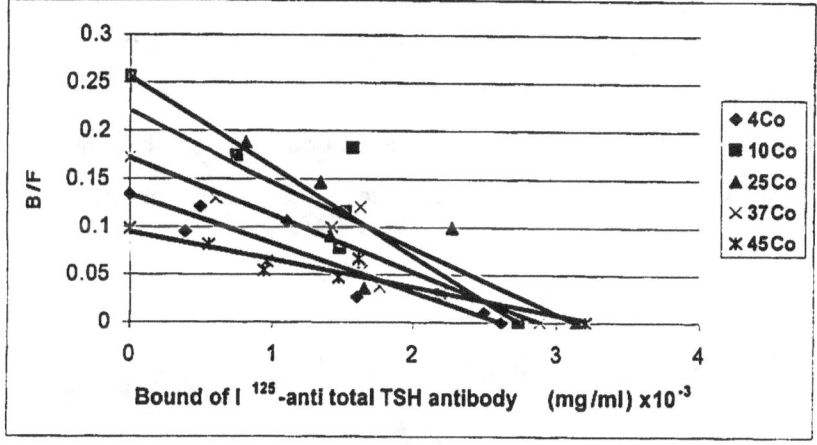

B Malignant

Figure (3.16): Scatchard plots of ^{125}I-anti TSH antibody binding with TSH in: A) Benign thyroid homogenate tissue at five different temperatures, B) Malignant thyroid homogenate tissue at five different temperatures. Details are described in section (2.7.3.2).

Time course data shown in Figure (3.15) fits the first order kinetics for the association of TSH with its specific Ab due to the biomolecularity of this reaction, the following equation [153]:

$$\ln(AbAg)_e = \frac{(Ab)_T - (AbAg)_t (AbAg)_e (Ag)_T}{(Ab)_T [(AbAg)_e - (AbAg)_t]} = k_{+1}t \left[\frac{(Ab)_T (Ag)_T - (AbAg)_e}{(AbAg)_e} \right] \quad (4)$$

Could be simplified to equation (5) in order to fit the data of the first order kinetics.

$$\ln \frac{(AbAg)_e}{(AbAg)_e - (AbAg)_t} = k_{+1}t(Ab)_T (Ag)_T (AbAg)_e \quad \ldots\ldots\ldots\ldots (5)$$

Where k_{+1} is the kinetic association constant, $(Ab)_T$ is the total concentration of the ^{125}I-anti tTSH Ab, $(Ag)_T$ is the total concentration of TSH in thyroid tissue homogenate, $(AbAg)_e$ is the concentration of (^{125}I-anti tTSH/TSH) complex formed at equilibrium and $(AbAg)_t$ is the total concentration of complex formed after time (t).

Since in some cases of our work the percent of binding was small and most of labeled Ab remained free and only small fraction of $(Ab)_T$ is bound even at equilibrium (pseudo-first order conditions), so that the following equation could be used in order to fit the data of first order kinetics:

$$\ln \frac{(AbAg)_e}{(AbAg)_e - (AbAg)_t} = t.k_{obs.} \quad \ldots\ldots\ldots\ldots\ldots\ldots\ldots (6)$$

Figure (3.17) shows that the plotting of $\ln \dfrac{(AbAg)_e}{(AbAg)_e - (AbAg)_t}$ against time gives a straight line with a slope equal to the observed value of first order rate constant ($k_{obs.}$) in min^{-1}, the association rate constant k_{+1} was calculated from the following formula:

$$k_{obs.} = k_{+1} \cdot \frac{(Ab)_T . (Ag)_T}{(AbAg)_e} \quad \ldots\ldots\ldots\ldots\ldots\ldots\ldots\ldots (7)$$

The half-life time of association ($t_{1/2}$)$_{ass.}$, which represents the time needed for the formation of half amount of the complex at equilibrium was determined from the concentration of the complex at equilibrium and the time course curve. While the half-life time of dissociation ($t_{1/2}$)$_{diss.}$, was determined from:

$$(t_{1/2})_{diss.} = \ln \frac{2}{k_{-1}} = \frac{0.693}{k_{-1}} \quad \ldots\ldots\ldots\ldots\ldots\ldots\ldots (8)$$

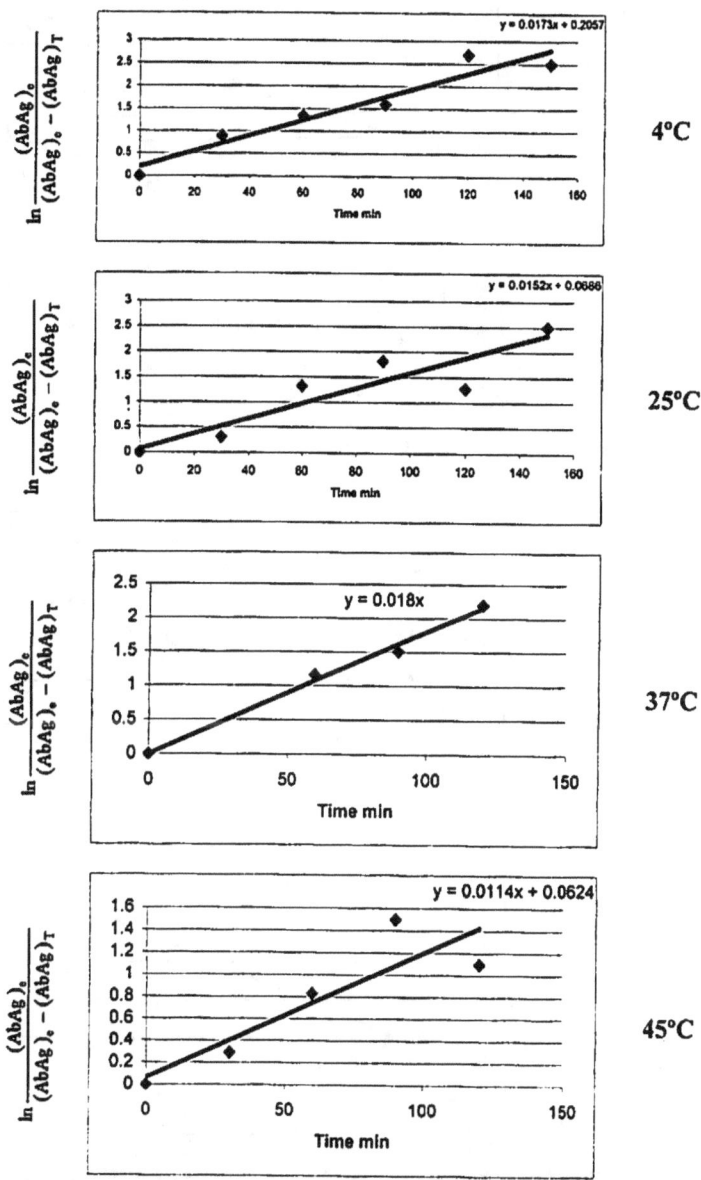

Figure: (3.17a). Kinetics of ^{125}**I-anti total TSH antibody binding with TSH in Benign thyroid Homogenate Tissue at four different Temperatures. Details are described in section (2.7.3.3).**

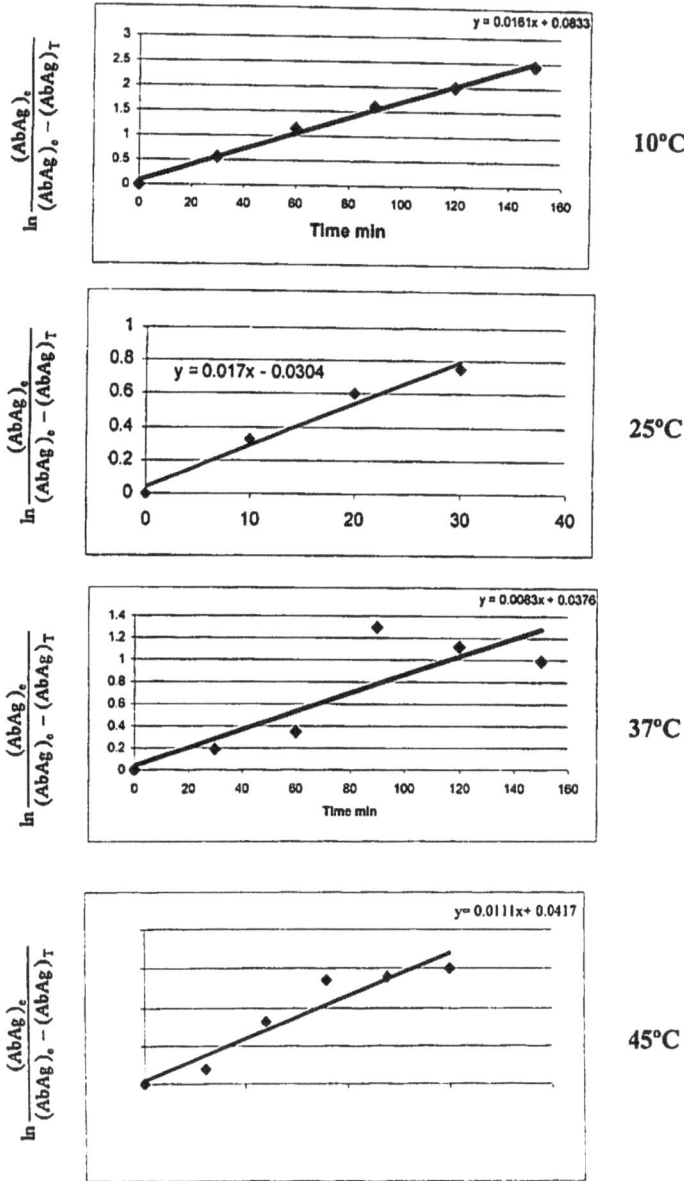

Figure: (3.17b). Kinetics of ^{125}I-anti total TSH antibody binding with TSH in malignant thyroid Homogenate Tissue at four different Temperatures. Details are described in section (2.7.3.3).

The values of k_{-1} were obtained from equation (3). Table (3.3) shows all these parameters ($k_{obs.}$, k_{+1}, k_{-1}, $t_{½\ ass.}$, $t_{½\ diss.}$) at different temperatures except

10°C and 4°C in benign and malignant thyroid tissue homogenates respectively because the maximum binding occurs at 30 min, therefore, the first order kinetic could not be applied.

Table (3.3): The effect of temperature on the kinetic parameters of ^{125}I-anti tTSH Ab binding to TSH in benign and malignant thyroid tissue homogenates. All other details are described in section (2.7.3.3).

Temp.	$k_{obs.}$ (min^{-1})		k_{+1} (mg^{-1}.ml.min^{-1})		k_{-1} (min^{-1})		$t_{½ ass.}$ (min)		$t_{½ diss.}$ (min)	
(°C)	B	M	B	M	B	M	B	M	B	M
4	0.017	-	94.4	-	2.052	-	22	-	0.33	-
10	-	0.016	-	69.5	-	0.74	-	30	-	0.943
25	0.015	0.017	76	60.7	1.26	0.89	19	20	0.55	0.77
37	0.018	0.0083	95.7	34.4	1.67	0.58	30	20	0.41	1.19
45	0.0114	0.0111	76	45.8	1.38	0.46	29	15	0.502	1.506

B : Benign, M: Malignant.

3.6.3. The thermodynamics studies of TSH binding in benign and malignant tissues with ^{125}I-anti TSH Ab

3.6.3.1. Thermodynamic parameters of standard state

The dependence of the equilibrium binding constant (affinity constant) for the binding of ^{125}I-anti TSH Ab to TSH in benign and malignant thyroid tissues on the temperatures can be observed from Van't Hoff plot [154], as shown in Figure (3.18). The results obtained from Van't Hof plot revealed that $\Delta H°$ in general has small values and their positive sign ascertain that the reaction was nearly endothermic. The small positive value of $\Delta H°$ may indicate a favorable interaction between ^{125}I-anti TSH Ab with TSH in benign and malignant homogenates. These include the non-covalent interaction, which are fundamentally electrostatic in nature such as charge-charge, charge-dipole, dipole-dipole, charge-induced dipole, dipole-induced dipole interactions and hydrogen bonds. The sum of these types of interactions can yield some stabilization to the folded structure of the complex.

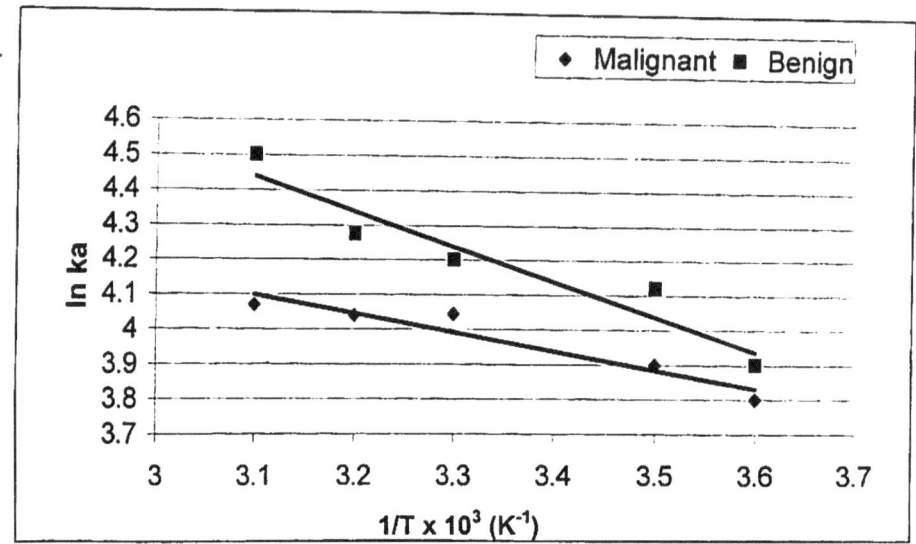

Figure (3.18): Van't Hoff plot for the ^{125}I-anti total TSH antibody binding with TSH in Benign and malignant thyroid homogenate tissues. Details are described in section (2.7.3.4).

Table (3.4) shows the values of thermodynamic parameters at standard state of TSH in benign (A) and malignant (B) tissues at five different temperatures. The negative values of $\Delta G°$ reflect the stability of complex and hence, the high affinity of the reactants. So, our system is characterized by the solute contribution of $\Delta S°$ to the stability of the complex formed, while $\Delta H°$ has a little or no effect.

Table (3.4): Thermodynamic parameters at standard state of ^{125}I-anti tTSH Ab binding with TSH in benign and malignant thyroid tissue homogenates. Details are described in section (2.7.3.4).

Temp. (°C)	$\Delta H°$ (KJ.mol^{-1})		$\Delta G°$ (KJ.mol^{-1})		$\Delta S°$ (J.mol^{-1}.K^{-1})	
	Benign	Malignant	Benign	Malignant	Benign	Malignant
4	3.751	5.481	-8.81	-9.05	45.34	52.45
10	3.751	5.481	-9.34	-10.58	46.25	56.75
25	3.751	5.481	-10.13	-10.45	46.58	53.45
37	3.751	5.481	-10.41	-10.48	45.68	51.48
45	3.751	5.481	-10.57	-11.89	45.03	54.62

High values of positive $\Delta S°$ suggest that the binding was entropically driven. Entropy was the driven force for the occurrence of the binding this indicates that the hydrophobic interactions played an important role in stabilizing complexes.

Moreover, the same table indicates that $\Delta H°$ and $\Delta S°$ values for malignant tumors homogenate were higher than that of benign tumors homogenate, because of the differences in the affinity constant between benign and malignant tumor homogenates.

3.6.3.2. Thermodynamic parameters of transition state

The transition state theory proposes that the association of two substances to form the final product proceeds through the formation of an activated complex (transition state). Consequently, the interaction of ^{125}I-anti TSH Ab with TSH in benign and malignant tissue homogenates could be represented as followed:

^{125}I-anti TSH Ab + TSH $\longrightarrow$ [^{125}I-anti TSH/TSH] $\longrightarrow$ ^{125}I-anti TSH Ab/TSH
an activated complex   (Product)
(Transition State)

The thermodynamic parameters of the transition state (ΔH^*, ΔS^* and ΔG^*) could be determined from Arrhenius equation and the kinetic constants.

Figure (3.19) shows the Arrhenius plots [155] plots of ln k_{+1} against $1/T$ values. The slope on the straight line represents the activation energy (Ea).

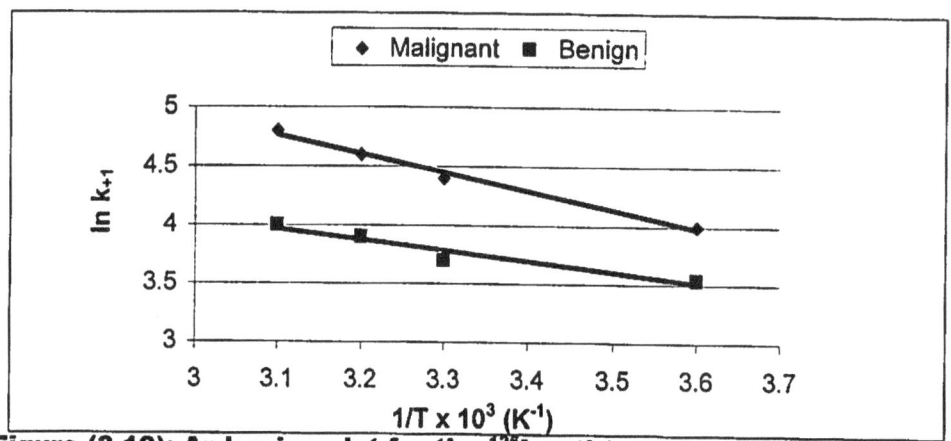

Figure (3.19): Arrhenius plot for the ^{125}I-anti total TSH antibody binding with TSH in Benign and malignant thyroid homogenate tissues. Details are described in section (2.7.3.4).

Table (3.5) shows the values of thermodynamic parameters of the transition state (E_a, ΔH^*, ΔS^* and ΔG^*). The high values of activation energy 4.98 KJ.mol^{-1} and 11.4 KJ.mol^{-1} of benign and malignant homogenate respectively which represents the required energy to overcome the energy barrier of the transition state of the formation of (^{125}I-anti TSH Ab/TSH) complex. Also the value of activation energy is in accordance with the high positive values of ΔG^*, which indicates that formation of the activated complex is a non-spontaneous process and required a lot of energy (equal to E_a) to overcome the transition state energy barrier and giving the final product, whereas the high negative ΔS^* revealed that the activated complex has more ordered structure than the reactants.

From the results obtained for the thermodynamic parameters in the transition state, it can be concluded that the positive values of ΔH^* and high positive value of ΔG^* are favorable to overcome the energy barrier of transition state, the negative values of ΔS^* shows more arranged structure for the activated complex, while the positive values of ΔG^* was mainly attributed to

85

decrease in the entropy of the transition state ($\Delta S^* < 0$). In addition the positive values of ΔH^* shows that the heat content of the activated complex was more than that of isolated species.

Table (3.5): Thermodynamic parameters at transition state of ^{125}I-anti tTSH Ab binding with TSH in benign and malignant thyroid tissue homogenates. Details are described in section (2.7.3.4).

Temp. (°C)	E_a B	E_a M	ΔH^* (KJ.mol^{-1}) B	ΔH^* (KJ.mol^{-1}) M	ΔG^* (KJ.mol^{-1}) B	ΔG^* (KJ.mol^{-1}) M	ΔS^* (J.mol^{-1}.K^{-1}) B	ΔS^* (J.mol^{-1}.K^{-1}) M
4	4.98	-	2.68	-	5.71	-	-196.4	-
10	-	11.4	-	9.04	-	59.2	-	-177.2
25	4.98	11.4	2.51	8.92	62.82	62.82	-200.3	180.8
37	4.98	11.4	2.41	8.82	64.27	66.9	-199.3	-187.3
45	4.98	11.4	2.34	8.75	66.6	67.9	-202	-186.0

B: benign, M: Malignant.

3.7. Isolation of (^{125}I-anti TSH Ab/TSH) Complex and Unbound of ^{125}I-anti TSH Ab in Benign and Malignant Thyroid Tissue Homogenates

Figure (3.20) shows the results of gel filtration technique used to separate (^{125}I-anti TSH Ab/TSH) complex from unbound ^{125}I-anti TSH Ab for benign and malignant thyroid tissues, respectively. In this experiment the amount of protein and radioactivity were measured for every fraction.

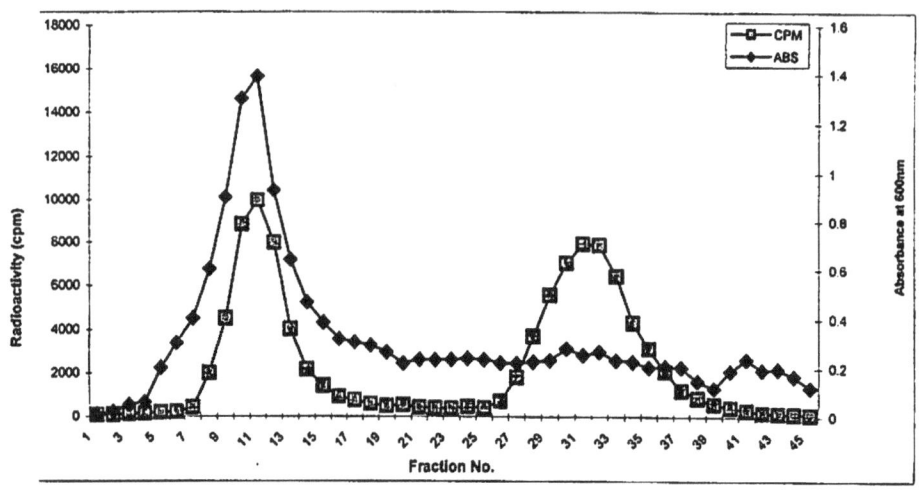

A: Benign

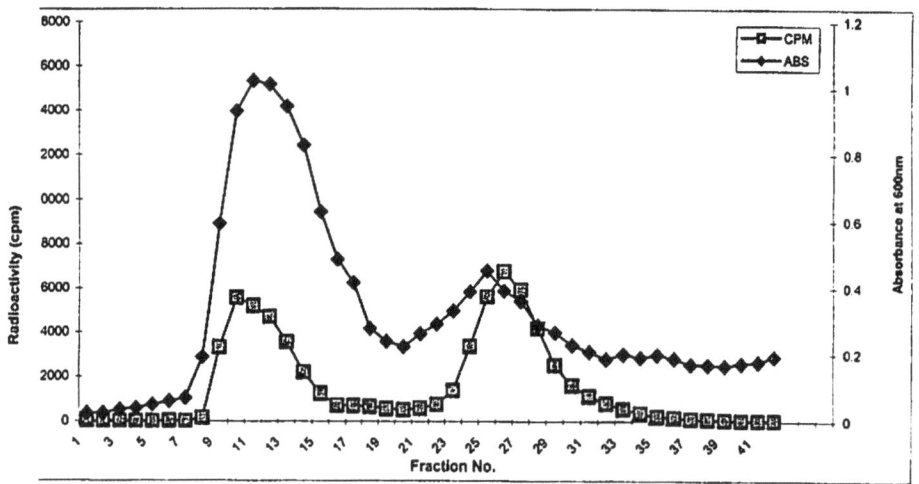

B: Malignant

Figure (3.20): The elution profile of ^{125}I-anti total TSH Ab binding with TSH in, A) Benign thyroid tissue homogenate, B) Malignant thyroid tissue homogenate from sephadex G-150 column. Details are described in section (2.8).

Regarding benign tissue homogenate, all trials of this experiment revealed two peaks of radioactivity. The first peak represented the complex with maximum radioactivity in fraction number (11) and the second peak represented the unbound (free) ^{125}I-anti TSH Ab with maximum radioactivity in fraction number (30), while two absorbance peaks for protein content were obtained with maximum absorbance in fraction number (11) for the complex and in fraction number (30) for the unbound (free) ^{125}I-anti TSH Ab as shown in Figure (3.20A).

With regard to tissue, all details of this experiment also revealed two peaks of radioactivity with maximum radioactivity in fraction (10 and 26) for the complex and the unbound ^{125}I-anti TSH Ab, respectively. Also, the absorbance peaks for protein content were obtained with maximum absorbance in fraction number (12 and 13) for complex and in fraction number (23 and 25) for unbound (free) ^{125}I-anti TSH Ab as shown in Figure (3.20B).

The isolation by gel filtration depends upon the difference of molecular weight (M.wt.) of the compounds and because the M.wt. of the complex was greater than that of the unbound ^{125}I-anti TSH Ab, so, the results indicate that the first peak was assigned for the complex and the second peak was assigned for the free. A previous study on thyroid gland indicated that the M.wt. of ^{125}I-anti TSH Ab/TSH complex was (180000 Dalton) [156].

3.8. Spectroscopic Studies on TSH/^{125}I-Anti TSH Ab Complex and Unbound ^{125}I-anti TSH Ab

3.8.1. The UV-spectrum of (^{125}I-anti TSH Ab/TSH) complex and Unbound ^{125}I-anti TSH Ab in benign and malignant thyroid tissue homogenates

Figure (3.21) illustrates the UV spectra of pooling fraction of (complex, unbound) at pH = 7.8. In addition, Table (3.6) shows the λ_{max} values of ^{125}I-anti tTSH/TSH Ab) complex and unbound ^{125}I-anti tTSH Ab in benign and malignant thyroid tissue homogenates. These (AbAg) complexes and unbound ^{125}I-Ab were isolated and obtained by gel filtration technique as described in section (2.9.1).

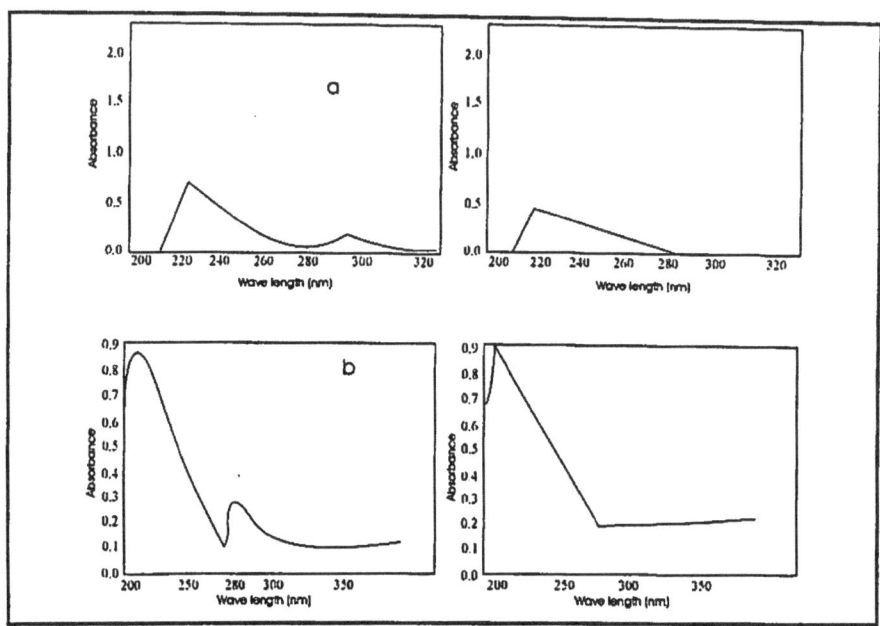

Figure (3.21). The UV-spectrum of (¹²⁵I-anti TSH Ab/TSH) complex and unbound ¹²⁵I-anti TSH Ab at neutral pH 7.8 in, a) benign thyroid tissue homogenate b) malignant thyroid tissue homogenate, (left: complex, right: unbound). Details are described in section (2.9.1).

Form Table (3.6) one can calculate the following:

- The spectrum shows that the λ_{max} for benign (TSH/anti tTSH) complex consists of two peaks. One at 224 nm and the other at 295 nm. As a result TSH has a characteristic spectrum and can be identified by its peaks, which were assigned to the peptide bonds [157] and tyrosine [158] residues respectively. It seems that tyrosine residues in TSH molecules was located in a way that part of it was on the surface of the protein molecule and the other part was buried, while the spectrum for the unbound ¹²⁵I-anti tTSH Ab shows that the λ_{max} consists of a single peak at 219.2 nm which assigned to the peptide bonds (tryptophan residues).

- On the other hand, the spectrum of the malignant thyroid tissue homogenate of (¹²⁵I-anti tTSH Ab/TSH) complex consists of two peaks one at 210 nm and the other at 283 nm, which assigned to the peptide bonds (histidine residues) and tyrosine residues respectively. The spectrum of the unbound ^{125}I-anti tTSH Ab consists of a peak at 208.5 nm which assigned to the peptide bonds (histidine residues).

Table (3.6): The λ_{max} (nm) values of UV-spectra of the (^{125}I-anti TSH Ab/TSH) complexes and unbound of ^{125}I-anti TSH Ab in benign and malignant thyroid tumors at pH 7.8 (all other details are described in section (2.9.1).

Source	Complex (^{125}I-anti TSH Ab/TSH)		Unbound ^{125}I-anti TSH Ab	
	λ_{max1}	λ_{max2}	λ_{max1}	λ_{max2}
Benign thyroid tissue homogenate	224	295	219	-
Malignant thyroid tissue homogenate	210	283	208.5	-

3.8.2. Factors affecting the absorption properties of (^{125}I-anti TSH Ab/TSH) complex and unbound of ^{125}I- anti TSH Ab in malignant thyroid tissue homogenate on UV spectrum

The absorption spectrum of a chromophore was primarily determined by the chemical structure of the molecule. However, a long number of environmental factors produces detectable change in λ_{max} and ε. Environmental factors such as pH and polarity of the solvent provide the basis for the use of absorption spectroscopy in characterizing molecules [157].

3.8.2.1. pH effect

The pH of the solvent determines the ionization state of the ionizable chromophore in the protein molecule. Table (3.7) and Figure (3.22) show the λ_{max} values of human TSH (complex and unbound) at different pH (4, 7.8 and 12). At pH (7.8) two λ_{max} were determined for complex and one λ_{max} for unbound of ^{125}I-anti tTSH Ab for malignant tumor homogenate. While at an acidic pH 4, two λ_{max} were obtained for malignant complex, the first one at 198 nm and the second at 284 nm that assigned to tyrosine and tryptophan [158] residues (Figure 3.22a). On the other hand, the malignant (unbound) gives only one peak at λ_{max} 198 nm that assigned to tyrosine (Figure 3.22b).

Table (3.7): The effect of pH on λ_{max} (nm) of the (^{125}I-anti TSH Ab/TSH) complex and unbound of ^{125}I-anti TSH Ab in malignant thyroid tissue homogenate. Details are described in section (2.9.2.1).

| pH | Malignant thyroid tissue homogenate | | | |
| | Complex (^{125}I-anti TSH Ab/TSH) | | Unbound ^{125}I-anti TSH Ab | |
	λ_{max1}	λ_{max2}	λ_{max1}	λ_{max2}
4	198	284	198	-
7.8	210	283	208	-
12	230	285	230	-

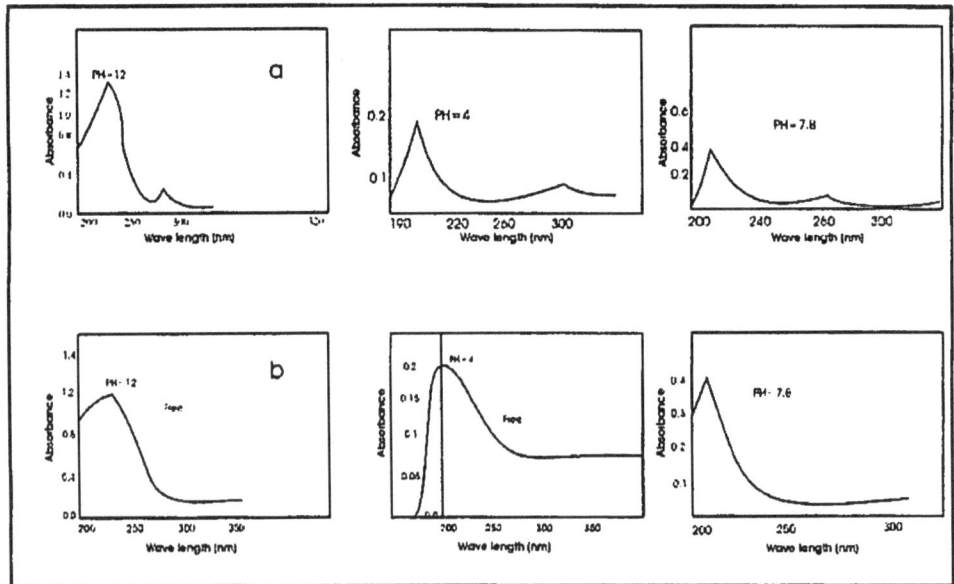

Figure (3.22). The pH effect on UV-spectrum of, a) (^{125}I-anti TSH Ab/TSH) complex, b) unbound ^{125}I-anti TSH Ab in malignant thyroid tissue homogenate at different pH (12, 4, 7.8). Details are described in section (2.9.2.1).

When the pH value was increased from 7.8 to 12, an increase in λ_{max1} and the absorbance while λ_{max2} remain constant. This increase in λ_{max1} was due to the dissociation of the phenolic (OH) of tyrosine residuee ($pK_a = 10.07$) giving an ionized form of this amino acid, which absorbed at higher wave length (red shift). The spectral shifts of protein produced by pH cannot be simply attributed to the inductive effect of vicinal charges; such spectral changes must therefore be attributed mainly to rearrangement of secondary and tertiary structure, although the possibility of field effects due to unusually close conjugation of charges to aromatic groups was not excluded.

3.8.2.2. The effect of solvent polarity

Proteins are rarely studied in completely non-polar solvents because most proteins were either insoluble or denaturated in such solvents [159]. However, significant solvent effects can be induced by the use of mixtures of water and substance of a different polarity such as; methanol, chloroform, glycerol, polyethylene glycol, KCl and urea.

The concentration of the solvent which was used in this experiment (20%) does not appear to produce conformational changes in most proteins so the perturbing solvent does not alter the conformational of the protein but alter the peak position and intensities by altering the energy and probability of electronic transitions by solvation of the chromophores which are in contact with the solvent or polarity buried chromophores.

- **The Effect of 20% Methanol on UV-Spectrum**

Table (3.8) shows the λ_{max} of all the solvents, and Figure (3.23A) shows the λ_{max} values of (anti tTSH/TSH Ab) complex and unbound ^{125}I-anti TSH Ab in 20% methanol at pH 7.8. The malignant (^{125}I-anti TSH/TSH Ab) complex has two λ_{max} one at 219 nm and the other 283 nm, which assigned to tryptophan and tyrosine residues. These was shift in λ_{max1} at 219 nm while λ_{max2} remain constant, this shift is called red shift towards longer wavelengths in methanol at 20% concentration. These shifts are attributed to $\pi = \pi^*$ transitions. On the other hand, the unbound ^{125}I-anti TSH Ab in Figure (3.23B) shows λ_{max} at 216 nm, which was assigned to tryptophan residues. So, there was a red shift towards longer wavelength due to $\pi = \pi^*$ transitions.

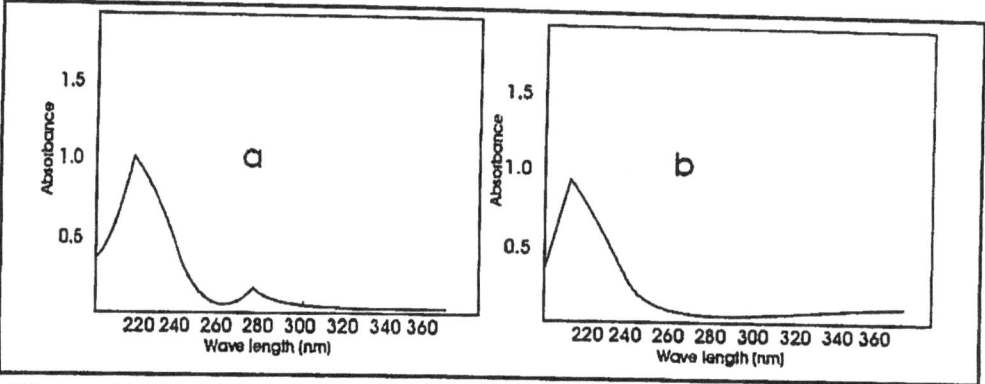

Figure (3.23). The effect of 20% methanol on the λ_{max} of UV-spectrum at pH 7.8 in, A) complex, b) free in malignant thyroid tissue homogenate. Details are described in section (2.9.2.2).

- **The Effect of 20% Chloroform**

    Table (3.8) shows the λ_{max} of this solvent at pH 7.8 in malignant thyroid tissue homogenate. The table shows that when protein was shifted to a less polar solvent (chloroform), the λ_{max1} = 206 nm which was assigned to phenylalanine residues decreased and the absorbance also in (anti tTSH/TSH Ab) complex with the absence of λ_{max2}. This shift toward a shorter wavelength was called blue shift and this shift was due to n $\rightarrow$ π^* transition as shown in Figure (3.24A). While the unbound ^{125}I-anti TSH Ab, there was no effect as illustrated in Figure (3.24B).

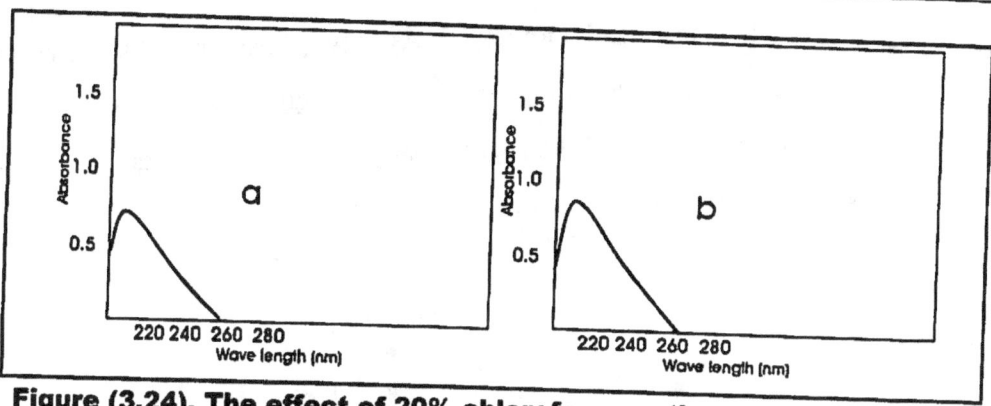

Figure (3.24). The effect of 20% chloroform on the λ_{max} **of UV-spectrum at pH 7.8 in, A) complex, b) free in malignant thyroid tissue homogenate. Details are described in section (2.9.2.2).**

- **The Effect of 20% Polyethylene Glycol and 20% Glycerol**

The effect of these solvent on the ^{125}I-anti TSH Ab/TSH complex and unbound of ^{125}I-anti TSH Ab on UV-spectrum were illustrated in Table (3.8). In addition, Figure (3.25A) shows the effect of 20% polyethylene glycol on (^{125}I-anti TSH Ab/TSH) complex. There was a slight red shift in λ_{max1} 211 nm which was assigned to histidine residues due to $\pi \rightarrow \pi^*$ transition with the absence of λ_{max2}. Moreover, Figure (3.25B) shows no effect for 20% polyethylene glycol on unbound ^{125}I-anti TSH Ab.

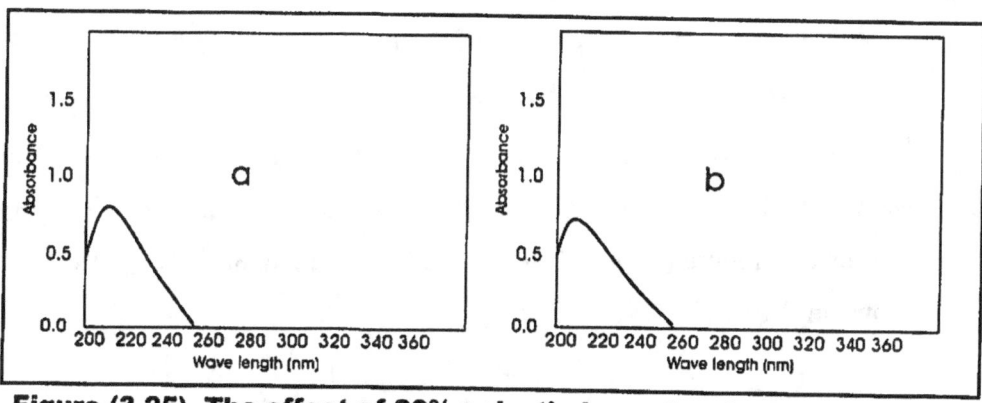

Figure (3.25). The effect of 20% polyethylene glycol on the λ_{max} **of UV-spectrum at pH 7.8 in, A) complex, b) free in malignant thyroid tissue homogenate. Details are described in section (2.9.2.2).**

On the other hand, Figure (3.26A) shows the effect of 20% glycerol of (^{125}I-anti TSH Ab/TSH) complex. Their was slight red shift at λ_{max1} 211nm due to $\pi \rightarrow \pi^*$ transition, while λ_{max2} not changed. In addition, Figure (3.26B) shows the effect of 20% glycerol on unbound ^{125}I-anti TSH Ab.

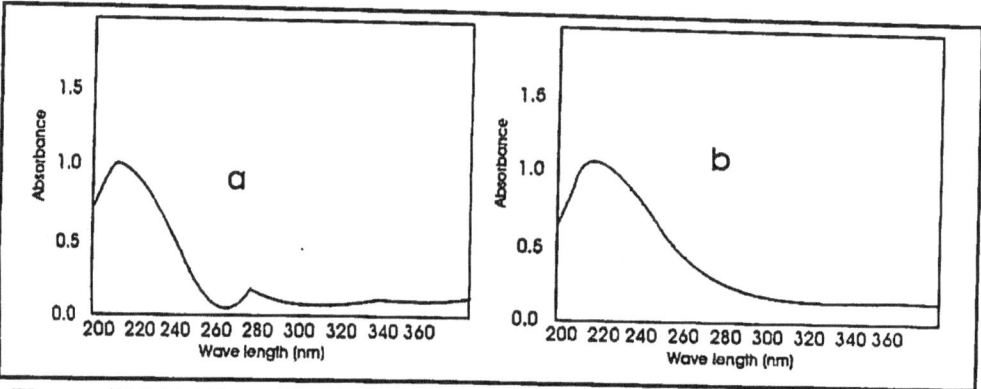

Figure (3.26). The effect of 20% glycerol on the λ_{max} of UV-spectrum at pH 7.8 in, A) complex, b) free in malignant thyroid tissue homogenate. Details are described in section (2.9.2.2).

There was a red shift towards longer wavelength at λ_{max1} 215 nm that assigned to tryptophan residues, while no peak of λ_{max2} was detected.

- **The Effect of 20% Urea and 20% KCl**

   Table (3.8) shows the effect of these solvents. In addition to that Figure (3.27) shows the effect of 20% urea on the ^{125}I-anti TSH Ab/TSH complex on UV-spectrum. The λ_{max2} remained constant while λ_{max1} was shifted to a longer wavelength 224 nm (tyrosine residues). These red shift due to $\pi \rightarrow \pi^*$ transition as shown in Figure (3.27A). The same effect of unbound ^{125}I-anti TSH Ab was shown in Figure (3.27B).

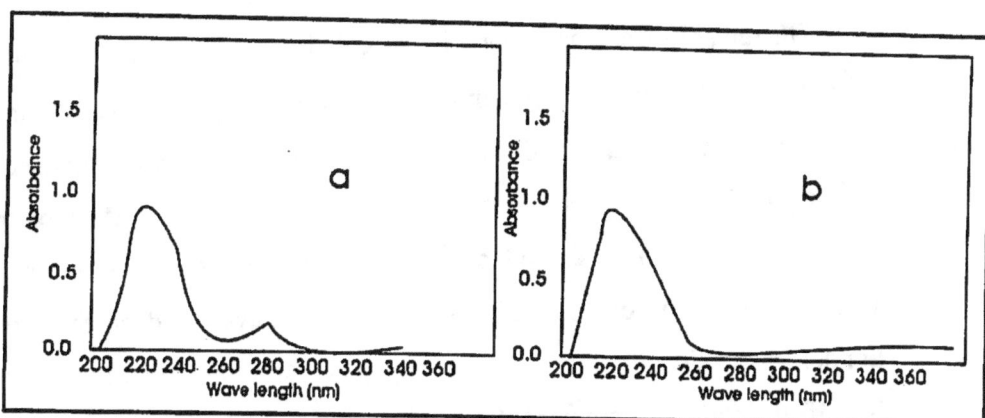

Figure (3.27). The effect of 20% urea on the λ_{max} of UV-spectrum at pH 7.8 in, A) complex, b) free in malignant thyroid tissue homogenate. Details are described in section (2.9.2.2).

On the other hand, Figure (3.28A) shows the effect of 20% KCl on the ^{125}I-anti TSH Ab/TSH complex. There was a longer red shift in λ_{max1} with the absence of λ_{max2}. The λ_{max1} (254 nm) which attributed to phenylalanine residues was due to $\pi \rightarrow \pi^*$ transition of the aromatic ring of phenylalanine residues. In addition Figure (3.28B) shows the same effect of 20% KCl on unbound ^{125}I-anti TSH Ab.

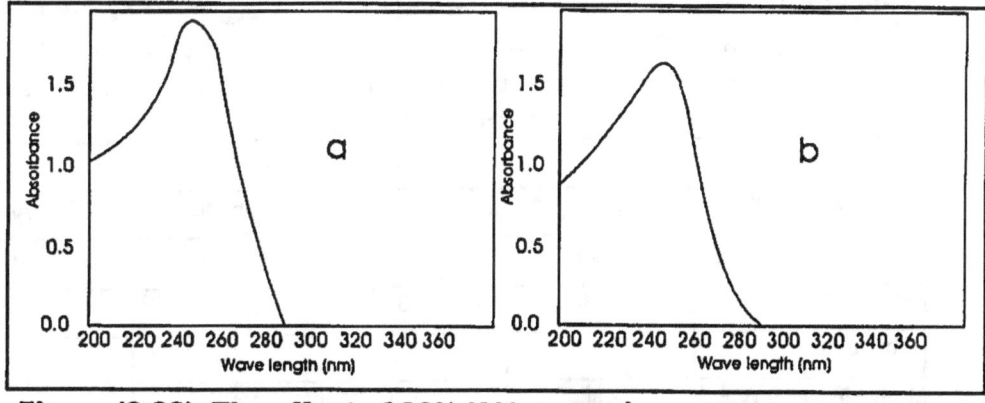

Figure (3.28). The effect of 20% KCl on the λ_{max} of UV-spectrum at pH 7.8 in, A) complex, b) free in malignant thyroid tissue homogenate. Details are described in section (2.9.2.2).

The red shift was due to negative or positive charges of the solvent anions and cations [158,159] which might interact directly with π electron system of the benzene ring in tyrosine residues or any other ionizable group exposed to the solvent, so, Table (3.8) shows that 20% KCl has marked effect on protein structure. The data in the same table appear to be as a result of a net interaction of the solvent with non polar chromophores that gave a red polarization shift (London dispersion interaction) and with polar chromophores that give a red shift if their dipole moment (or hydrogen binding to solvent molecules) increases in excited state [160], but a blue shift if their dipole moment (or hydrogen binding to solvent molecules) decreases in excited state. So, the shift was due to the alteration of the energy and probability of electronic transition by the solvent, by polarization effect or by changes in permanent dipole moment during excitation, i.e., the dipole hydrogen binding, which will tend to produce either a short or a long wave shift depending on nature of the electronic transition and whether the solute is the hydrogen donor or hydrogen acceptor [161,162].

Table (3.8): The effect of different solvents on λ_{max} (nm) of the (^{125}I-anti TSH Ab/TSH) complex and unbound of ^{125}I-anti TSH Ab in malignant thyroid tissue homogenate. Details are described in section (2.9.2.2).

20% Solvent	Malignant thyroid tissue homogenate			
	Complex (^{125}I-anti TSH Ab/TSH)		Unbound ^{125}I-anti TSH Ab	
	λ_{max1}	λ_{max2}	λ_{max1}	λ_{max2}
Methanol	219	283	216	-
Chloroform	206	-	210	-
Polyethylene glycol	211	-	210	-
Glycerol	212	283	215	-
Urea	224	282	224	-
KCl	254	-	252	-

3.8.3. The effect of NaCl concentration on thermal stability of ^{125}I-anti TSH Ab/TSH complex and unbound ^{125}I-anti TSH Ab in malignant thyroid tissue homogenate

Figure (3.29) and (3.30) show the unfolding of ^{125}I-anti TSH Ab/TSH complex and unbound ^{125}I-anti TSH Ab respectively as a function of temperature by using NaCl (0.1 M and 0.01M) at two λ_{max} (292 and 295 nm) for tryptophan and tyrosine residues respectively. The absorbance was increased with increasing the temperature because a buried chromophore [143] becomes exposed to the solvent during denaturation; so, the conformational transition for the complex (^{125}I-anti TSH Ab/TSH) and unbound ^{125}I-anti TSH Ab was confirmed by the hyperchromicity.

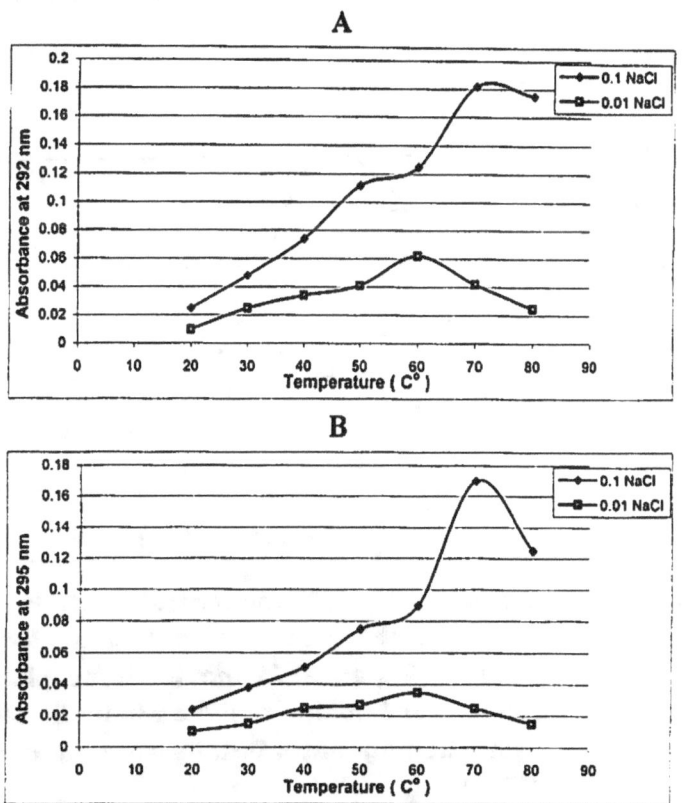

Figure (3.29). The effect of NaCl concentration (0.01 M, 0.1M) on the thermal stability of the (^{125}I-anti total TSH/TSH Ab) complex at, A) 292nm for tryptophan residues, B) 295 nm for tyrosine residues in malignant thyroid tissue homogenate. Details are described in section (2.9.3).

A

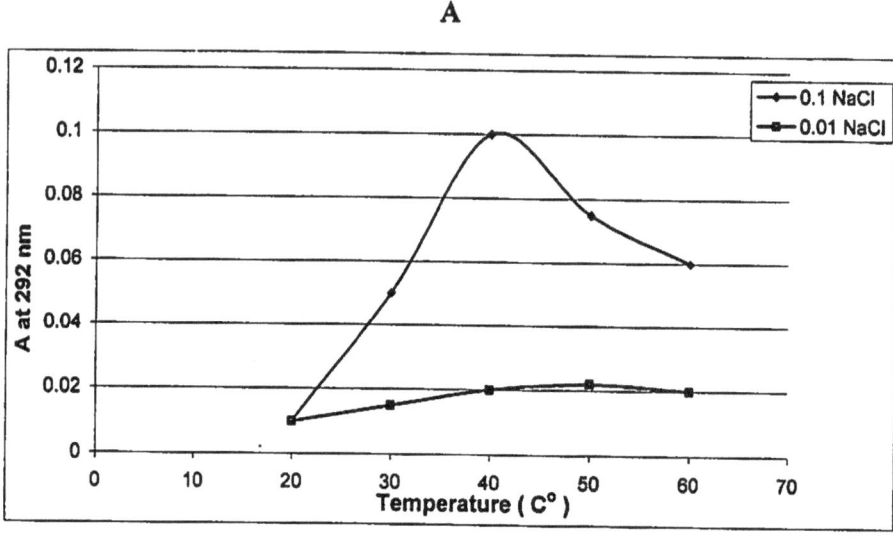

B

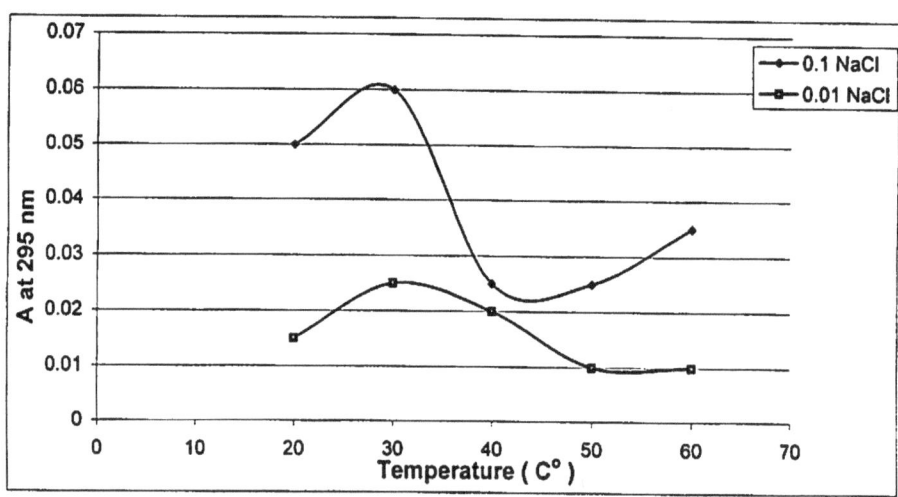

**Figure (3.30). The effect of NaCl concentration (0.01 M, 0.1M) on the
thermal stability of the unbound of ^{125}I-anti TSH
antibody Ab, at, A) 292nm for tryptophan residues, B)
295 nm for tyrosine residues in malignant thyroid
tissue homogenate. Details are described in section
(2.9.3).**

Figure (3.29) shows the ^{125}I-anti TSH Ab/TSH complex was more stable at high NaCl concentration (0.1M). Since high temperature (70°C) was needed for completely unfolding. The complete unfolding in a 0.01M NaCl happened at 60°C. While Figure (3.30) shows that unbound ^{125}I-anti TSH Ab was more stable at higher NaCl concentration (0.1 M). A high temperature 40°C was needed at 292 nm for complete unfolding and 30°C was needed at 295 nm for completely unfolding, while at 0.01 M NaCl no change happened.

Conclusion

- The results obtained show a significant increase of T_3 and T_4 levels and a significant decrease in TSH level in sera of thyrotoxicosis patients while no significant differences were observed in sera of multinodular goitre and thyroid carcinoma patients.

- The developed protocol for the assay of TSH in serum is suitable for the assessment of TSH in tissue.

- The kinetic studies of ^{125}I-anti total TSH antibody with TSH in benign and malignant tissue homogenate showed that the reaction is temperature and time dependent. The binding data fits pseudo-first order kinetics at (4, 25, 37, and 45°C) for benign and (10, 25, 37, and 45°C) for malignant thyroid tissue homogenates.

- The results obtained from the thermodynamic studies on the association of ^{125}I-anti TSH Ab with TSH in benign and malignant thyroid tissue homogenates indicate that the binding reaction occurs spontaneously $\Delta G° < 0$, and entropically–driven since $\Delta S° > 0$.

- Thermodynamic parameters at transition state and excited state may give idea about the nature and role of hydrophobic forces in stabilization of the complex formed.

- The spectroscopic studies revealed that the (^{125}I-anti total TSH Ab/TSH) complex and unbound ^{125}I-anti TSH Ab in benign and malignant thyroid tissues have a characteristics spectrum.

- Fractionation of benign and malignant thyroid tissues homogenate by gel filtration technique revealed the presence of two peaks, one for complex (^{125}I-anti TSHAb/TSH) and the second for unbound ^{125}I-anti TSH Abs.

Future Work

- The application of the development method of IRMA for the assessment of TSH in other tumor tissues.

- Determination of total TSH conc. In sera and thyroid tissues of patients with different thyroid carcinoma stages, and eventually comparison could be made among these various stages.

- Implementation of further studies of TSH-binding inhibitors with evaluation of the correlation of these compounds with the incidence of different groups of thyroid tumors.

- Infrared, nuclear magnetic resonance spectrometry studies on purified TSH in thyroid carcinoma tissues.

References

References

1. Greenspan F.S., and Gardner D.G.: Basic and Clinical Endocrinology; 6th ed.; Mc Grow Hill; 2001;P.202.

2. Russo D., Arturi F., and Wicker R.: J.Clin. Endocrinol. Metab.; 1995;80;1347.

3. Chao T.C., Jeng L.B., Lin J.D., and Chen M.F.:Journal of Sugery;1997;21;644.

4. Venkatraman L., Maxwell P., and Mc Gluggage W.: Journal of Clinical Pathology; 2001;54;314.

5. Junqueria L., Carneiro J., and Kelley R.:Basic Histology; 7th ed.; McGraw Hill; 1992;P.527.

6. Braverman L.,and Utiger R.:Werner and Ingbar's The Thyroid; A Fundemental and Clinical Textbook; 8th ed.;Lippincot; 2000; P.269.

7. Margaret S.:Gene Enginnering In Endocrinology; Humana Press. Inc.; 2000; P.99.

8. Burtis C., and Ashwodd E.:Tietz Textbook Of Clinical Chemistry; 3rd ed.; W.B. Saunders; 1999; P.1496.

9. Murray R. Granner D., Mayes P., and Victor R.: Harper's Biochemistry; 25th ed.; Mc Graw Hill; 2000; P.561.

10. Zilva J., Pannall R., and Mayne D.:Clinical Chemistry In Diagnosis And Treatment; 5th ed.; 1988; P.158.

11. Guyton A., and Hail J.: Medical Physiology; 9th ed.; W.B. Saunders; 1997; P.123.

12. Thomas E., Charles C., Robert C., and Joseph L.: Cecil Essentials of Medicine; 5th ed.; W.B. Saunders; 2001; P.555.

13. Wilson J., and Foster D.: Williams Textbook Of Endocrinology; 8th ed.; W.B. Saunders; 1992; P.357.

14. Brent G.: New England Jaurnal Of Medicine; 1994; 331; 847.

15. Motomura K., and Brent G.: Endocrine Metabolism Clin North America; 1998; 27; 1.

16. Cristopher H., Chilver E., Hunter J., and Boon N.: Davidson's principle And Practice Of Medicine; 18th ed.; Churchill Livingstone; 1999; P.559.

17. Orlo H., and Duj Q.: Textbook Of Endocrine Surgery; 2nd ed.; 1997; P.34.

18. Balint K.: Endocrine Physiology; 3rd ed.; McGraw Hill; 2001; P.307.

19. De Vita T., Hellman J., and Rosenberg A.: Cancer Principle And Practice Of Oncology; 5th ed.; Lippincott-Raven; 1997; P.1629.

20. Wenig M., Thompson L., Adair F., Shmookler B., and Heffess C.: Cancer; 1998;82; 740.

21. Raghavan D., Brecher M., Johnson D., Meropl N., Moots P., and Thigpen J.: Textbook Of Uncommon Cancer; 2nd ed.; John Wily and Sons: 1999; P.260.

22. Takashima S., Fukuda H., and Kobayashi T.:Jaurnal Of Clinical Ultrasound; 1994; 22; 535.

23. Wilson J., Foster D., Kronenberg H., and Larson P.: Williams Textbook Of Endocrinology; 9th ed.; 1998; P.136, 453, 482, 485.

24. Frank D., William C., Morris M., and Charles R.: Cancer; 1997; 79; 564.

25. Field B., Bloom G., Margaret C., Mary E., Larson R., Kotani M., Kariya T., and Dekker A.: Endocrinol. Metabol.; 1978; 47; 1052.

26. Rubin P., and Williams J.: Clinical Oncology; A Multidisciplinary Approach For Physicians and Students; 8th ed.; W.B. Saunders; 2001; P.648.

27. Mazzaferri E.: New England Jaurnal of Medicine; 1993; 328; 553.

28. Vanderpump M., Alexander L., and Scarpello J.: Clinical Endocrinology; 1998; 48; 419.

29. Robbins J., Merino J., Boice J., Ron E., Ain B., and Alexander R.: Ann. Int. Med.; 1991; 115; 133.

30. Mazzafferri E.: American Journal Of Medicine; 1992; 93; 359.

31. Parkin D., Pisani P., and Ferlay J.: CA Cancer J. Clin.; 1999; 49; 33.

32. Greenler R., Taylor M., and Bolden S.: CA Cancer, J.Clin.: 2000; 50; 7.

33. Giovanni L., Giovanni V., Michele C., Maria F., and Bianco A.: Lancet; 1999; 353; 637.

34. Paul G., Thomas S., and Leigh D.: J. Am. Coll. Surg.; 1999; 189; 253.

35. Hundahl S., Fleming D., and Fremgen M.: Cancer; 1998; 83; 2638.

36. Ebihara S., and Saikawa M.: Thyroidol Clin Exp.; 1998; 10; 89.

37. Gunderson L., and Tepper J.: Clinical radiation Oncology; Churchill Livingstone: 3rd ed.; 2000; P.534.

38. Tan G., and Gharib H.: Ann. Intern. Med.; 1997; 126; 226.

39. Shore R.: Radiation Res.; 1992; 131; 98.

40. Clifford C., Perez A., and Brady L.: Radiation Oncology management Decisions; Lippincott-Raven; 1999; P.297.

41. Schlumberger M., and and Pacini F.: Thyroid Tumors; Edition Nucleon; 1999; 320.

42. Kazakov V., Demidchik E., and Astakhova L.; Nature; 1992; 359; 21.

43. Schaller R., and Stevenson J.: Cancer; 1966; 19; 1063.

44. Michael P., Pinedo H., and Veronesi V: Oxford Textbook Of Oncology; 2nd ed.; Oxford University Press Inc.; 1995;P.2097.

45. Williams E., Doniach I., Bjarnason P. and Mickle W.: Cancer; 1977; 39; 215.

46. Goldger D., Easton F., and Canon L.: J. Natl. cancer Inst.; 1994; 86; 1600.

47. Landis S., Murray T., and Bolden S.: CA cancer J.Clin.; 1999; 49; 8.

48. Hedinger C., Williams E., and Sobin L.: Cancer; 1989; 63; 908.

49. Correa P., and Chen V.: Cancer; 1995; 75; 338.

50. Rossing M., Voigt F., Wicklund K., and Daling J.: Am. J. Epiemiol.; 2000; 151; 765.

51. Rosali J., carcangiu M., and De Lellis R.: Atlas of Tumor Pathology; 3rd ed.; Churchill Livingstone; 1993; P.28.

52. Volante M., Papotti M., Roth J., Ernst J., and Gianni B.: Amercal Journal of Pathology; 1999; 155; 1499.

53. Li Volsi A., and Bronner P.: Sugical Pathology; 1988; 1; 37.

54. Jore A., Housini I., Frank V., and Synder W.: Cancer; 1997; 80; 1110.

55. Sakamoto A., Kasai N., and Sugano H.: Cancer; 1983; 52; 1849.

56. Ain K.: Thyroid; 1998; 8; 715.

57. Eng C., Calyton D., and Schuffenecker: J. Clin Endocrinol Metab.; 1995; 80; 2577.

58. Li Volsi A., Brooks J., and Arendash B.: American Jaurnal Clinical Pathology; 1987; 87; 434.

59. O'Riordain D., O'Brien T., Weaver A., Gharib H., and Hay I.: Surgery; 1994; 116; 1017.

60. Snow K.: Endcrinol. Metab. Clin. North Am.; 1994; 23; 157.

61. Sobin L., and Wittekind C.: TNM Classification Of Malignant Tumors; 5th ed.; Wiley-Liss; 1997; P.214.

62. Brierly J., Panzarella T., and Tsang R.: Cancer; 1997; 79; 2414.

63. Halland J., and Bast C.: Cancer Medicine; 14th ed.; Williams and Wilkins; 1997; P. 2483.

64. Yamashita H., Noguchi S., Murakami N., Kawamoto H., and Shin W.: Cancer; 1997; 80; 2268.

65. Shaha A., Loree T., and Shaha J.: Surgery; 1995; 118; 1131.

66. Gautvik K., Talle K., and Hafer B.: Cancer; 1989; 63; 175.

67. Tennvall J., Lundell G., Hallquist A., Walling G., and Tallroth: Cancer; 1994; 74; 1348.

68. Lawrence M., Stephen J., and Maxine P.: Current Medical Diagnosis and Treatment; 40th ed.; McGraw Hill; 2001; P.1106.

69. Jarlow A., and Hegedus E.: Journal Internal Medicine; 1991; 229; 159.

70. Venkatesh Y., Ordonez N., Schultz P., and Simpson W.: Cancer; 1990; 66; 321.

71. Ruegemer J., Hay I., and Bergstralh H.:J. Clin Endocrinol Metab.; 1998; 63; 960.

72. Moley F., Pyke C., Hay I., and Goellner J.: Sugery; 1991; 110; 964.

73. Dayan C., and Colin M.: Lancet; 2001; 357; 619.

74. Brunt L., and Wells S.: Endocrine Surgery; 1987; 67; 263.

75. Pineda J., Lee T., Ain K., Reynods J., and Robbins j.: Radiology; 1998; 161; 352.

76. Shimamoto K., Endo T., and Ishigaki T.: J. Ultrasound Med.; 1993; 12; 673.

77. Schneider A., Ron E., Lubin J., Stovall M., and Gierlowski T.: J. Clin Endocrinol Metab.; 1993; 77; 362.

78. Lees W., Vahl S., Watson L., and Russell R.: British Journal Of Surgery; 1987; 65; 681.

79. Tramalloni J., Leger A., Correas J., Helenon O., and Moreau F.: J., Radiol.; 1999; 80 (3); 271.

80. Charib H.: Myo Clin Proc.; 1994; 69; 44.

81. Takashima S., Fukuda H., and Koboyashi T.: J. Clin. Ultrasound; 1994; 22; 535.

82. Ozata M., Suzuki S., and Miyamato T.: J. Clin. Endocrinol. Metab.; 1994; 79; 98.

83. Spencer A., and Wang C.: Endocrinol. Metab. Clin. North Am.; 1995; 24; 841.

84. Takazawa M., Endo T., Erneux C., and Onaya T: Journal Of Endocrinology; 1995; 144; 561.

85. Loree T.: Semin. Surg. Oncol.; 1995;11;246.

86. Dorizi R., and Marin M.: Clinical Chemistry; 2001; 47 (6); 6.

87. Sizemore G.W.: Semin Oncol.; 1987; 14; 306.

88. Tisell L., Dilley W., and Wells S.: Thyroid; 1996; 6; 305.

89. Edington H., Watson C., Levine G., Tauxe N., and Kowal C.: Surgery; 1988; 104; 1004.

90. Krauze Y., Ish-Shalom S., De Jong R., Shibley N., and Lapidol M.: Clin. Nucl. Med.; 1994; 19; 416.

91. Gagel R., Goepfert H., and Callender D.: CA Cancer; J. Clin; 1996; 46; 261.

92. Mazzaferri E., and Jhiang S.: American Journal Of Medicine; 1994; 97; 418.

93. De Groot L., Kaplan E., and Mc Cromick M.: J Clin. Endocrinol. Metab.; 1990;71;414.

94. Takahashi H., Jiang N., Colum A., and Lee C.: J. Clin. Endocrinol. Metab.; 1978; 47; 870.

95. Wong J., Kaplan M., and Meyer K.: Endocrinol. Metab. Clin. North Am.; 1990; 19; 741.

96. Maxon H., Englaro E., Thomas S., Valimaki M., and Jhonston G.: J. Nucl. Med.; 1992; 33; 1132.

97. Sarne D., and Scheider A.: Endocrinol. Metab. North Am.; 1996; 25; 181.

98. Matsuzuka F., Migauchi A., and Katayama S.: Thyroid; 1993; 3;93.

99. Ekman E., Lunell G., and Tannvall J.: Otolaryngol Clin. Norht Am.; 1990; 23; 523.

100. Olson J., Hughes J., and Alpern H.: Surgery; 1992; 112; 1074.

101. Robert E., Peter A., Daryl K., and Victor R.: Harper's Biochemeistry; 22nd ed.; Appleton and Lange; 1993; P.478.

102. Larson P.: New England Jaurnal Of Medicine; 1982; 396; 23.

103. Ganong W.: Review Of Medical Physiology; 17th ed.; Appleton and Lange; 1995; P.290.

104. Christgau S.: Clinical Chemistry; 2000; 46(3); 313.

105. Mc Dermott T.: Endocrine Secrete; 2nd ed.; Hanley and Belfus; 1998; P.199.

106. Mazzaferri E.: Textbook Of Endocrinology; Elsevier; 1986; P.89.

107. Wartofsky L.: Principles and Practice Of Endocrinology and Metabolism; 2nd ed.; Lippincott; 1995; P.122.

108. Hay I., Bayer M., Kaplan M., Klee G., and Reed P.: Clinical Chemistry; 1991; 37; 2002.

109. Laurence D., Bennett P., Brown M.: Clinical Pharmacology; 8th ed.; Churchill Livingstone; 1977; P.303.

110. Philip F., Baxter D., Broadus A., and Laurence A.: Endocrinology and Metabolism; 2nd ed.; 1998; P.201.

111. Braveman L.: Diseases Of Thyroid; Humana Press; 1997; P.226.

112. Lee G., and Claud J.: Cecil textbook Of Medicine; 21st ed.; W.B. Saunders; 2000; P.1208.

113. Rapaport R.: Mol. Endocrinol; 1992;6;125.

114. Yukihiko W., Tada H., Hidaka Y., and Nobuyuki A.: Clinical Chemistry; 1999; 45(3); 407.

115. Rapaport R.: Endocr. Rev.; 1998; 19; 678.

116. Jameson J.: Thyroid; 1994; 4; 485.

117. Russo D., Chazenbalk G., and nagayama Y.: J. Biol. Chem.; 1990; 265; 20970.

118. Parmentier M., Libert F., and Maenjhaut C.: Science; 1989; 246; 1620.

119. Porcellini A., Ciullo I., Laviola L., Parma J., and Duprez L.: J. Clin Endocrinol. Metab.; 1994; 79; 657.

120. Gray C., and James V.: Hormones In Blood; 3rd ed.; Churchill Livingstone; 1979; P.499.

121. Takuda Y., Yasagi K., and Linda Y.: Clinical Chemistry; 1988; 34; 2360.

122. Peterson V., Smith B., and Hall R., J. Clin Endocrinol. Metab., 1975; 41; 199.

123. Micheal M.: Clinical Chemistry; 1999; 45(13); 1377.

124. Michael B., Janet L., and Edward P.: Clinical Chemistry Principles, Procedures, and Correlations; 4th ed.; 2000; P.375.

125. Nicoloff J., and Spencer C.: J. Clin Endocrinol. Metab.; 1990; 71; 553.

126. Franken N., Lenz H., and Maier L.: Clinical Chemistry; 1991; 37; 1035.

127. Rugg J., Flaa C., and Dawson S.: Clinical Chemistry; 1988; 34; 118.

128. Sotorrio P., Quiros A., and Jose M.: Clinical Chemistry; 1997; 43(12); 2428.

129. Stanley P., and Kricka L.: Bioluminescence and Chemoluminescence, Currnt Status; Wiley; 1991; P.123.

130. Donald L., and Robert W.: American Journal Of Medicine; 1998; 105; 524.

131. Warits R.; Steinberg M., Kinoshita F., Kelly C., and Richter R.: Regul. Toxocol. Pharmacol.: 1996; 24(2); 184.

132. Schlumberger M.: Clin. Endocrinol Metab.; 1986; 63; 960.

133. Dale E., Ahima R., Boers M., Joel E., and Fredric W.: Journal Of Clinical Investigation; 2001; 107(8); 1017.

134. Goldman J., Line B., Aamodt R., and Robbins J.:J. Clin. Endocrinol Metab.; 1980; 50; 734.

135. Schlumberger M.: Acta Endocrinologica; 1981; 98; 215.

136. Roger P., Taton M., van Sande J., and Dumont J.: J. Clin. Endocrinol. Metab.; 1988; 66;1158.

137. Filetti S.: New England Journal Medicine; 1988; 318; 753.

138. Kaplan L., and Pesce A.: Clinical Chemistry; 2nd ed.; C.V. Mosby; 1989; P.255.

139. Lowry O., Rosebergh N., Farr L., and Ronall J.: J. Biol. Chem.; 1951; 193; 265.

140. Willis M., Howell S., and Taylor K.: The Biochemistry Of The Polypeptide Hormones; 2nd ed.; John Wiley and Sons; 1985; P.17.

141. Rosalyn S.: Methods In Radioimmuno assay Of Peptide Hoemones; 1st ed.; Churchill Livingstone; 1976; P.1.

142. Scatchard G.: An N.Y. Acad. Sci.; 1949; 51; 660.

143. Freifelder D.: Physical Biochemistry; Application to Biochemistry and Molecular Biology; 2nd ed.; W.H. Freeman and Company; 1982; P.238.

144. Klein I., and Kaile O.: New England Jaurnal Of Medicine; 2001; 344(7); 501.

145. Pawetczyk T., Pawlikowski M., and Kunert J.: Jaurnal Of Endocrinology; 1996; 148; 193.

146. Christian H., and Torrens J.: 1999; 341(26); 2015.

147. Braunwald E., Issehbacher K., Petersdorf R., Wilson J., Martin J., Fauci A., and Kasper L.: Harrison's Principles Of Internal Medicine; 14th ed.; McGrow Hill; 1999; P.1999.

148. Kannon C.: Essential Endocrinology; A Primer For Non-Specialists; Plenum Medical Book; 1986; P.124.

149. Rosai J.: Ackerman's Surgical Pathology; 8th ed.; C.V. Mosby; 1996; P.391.

150. Kissane J.: Anderson's Pathology; 9th ed.; C.V. Mosby; 1999; P.1399.

151. Robert K., Peter A., Daryl K., and Victor R.: Harper's Biochemistry; 22nd ed.; Appleton and Lange; 1990; P.591.

152. Haro L., and Talaments F.: Mol. Cell Endocrinol.; 1985; 43; 199.

153. Seeley D., Wong Y., and Salhaniack H.: Biochem. Biophys. Acta; 1980; 632(4); 536.

154. Sanborn B., Anderson T., and Rechert L.: Biochemistry; 1987; 26; 819.

155. Ross P.D. and Subramanian T., Biochemistry; 1981; 20; 3096.

156. Birnbaumer L., and O'Malley B.: Methods in Enzymology; Hormone Action; Vol. 109; Academic Press. Inc.; 1985; P.677.

157. Freifeldar D.: Physical Biochemistry; 2nd ed.; John Wiely and Sons; 1982; P.500.

158. Yanari S., and Bovery F.: J. Biol. Chem.; 1960; 235(10); 2818.

159. Leach S., and Scherage H.: J. Biol. Chem.; 1960; 235(10); 2827.

160. Leach J.: Physical Principles and Technique Of Protein Chemistry; Part A; Academic Press Inc.; 1969; P.101.

161. Nagacura S., and Baba H.: J. Am. Chem. Soc.; 1952; 74; 5693.

162. Brealy G., and Raska M.: J. Am. Chem. Soc.; 1955; 77; 4462.

www.ingramcontent.com/pod-product-compliance
Lightning Source LLC
Chambersburg PA
CBHW080815180526
45168CB00006B/2455